# Airships and Balloons

*Other Piccolo True Adventures*

**Aidan Chambers**
*Haunted Houses*
*More Haunted Houses*
*Great British Ghosts*
*Great Ghosts of the World*

**Richard Garrett**
*Hoaxes and Swindles*
*True Tales of Detection*
*Narrow Squeaks!*
*Great Sea Mysteries*

**Frank Hatherley**
*Bushrangers Bold!*

**Marie Herbert**
*Great Polar Adventures*

**Carey Miller**
*Submarines!*

**Nicholas Monsarrat**
*The Boys' Book of the Sea*

**Sorche Nic Leodhas**
*Scottish Ghosts*

Piccolo True Adventures

# AIRSHIPS AND BALLOONS

Text illustrations by Chris Evans
Cover illustration by Peter Gregory

Carey Miller

**A Piccolo Original**
Pan Books London and Sydney

First published 1973 by Pan Books Ltd,
Cavaye Place, London SW10 9PG
3rd printing 1976
© Carey Miller 1973
ISBN 0 330 23471 4

Made and printed in Great Britain by
Cox & Wyman Ltd, London, Reading and Fakenham

# CONTENTS

1   A History of Airship Flight                         1

2   The Man Who Refused to be Beaten                    35

3   The Incredible Count von Zeppelin                   48

4   Reggie Warneford – First of the Giantkillers        62

5   Mathy and the Zeppelin Raiders                      74

6   Horror on the Ice Pack                              88

7   Journey into Disaster                              104

Chapter One

# A HISTORY OF
# AIRSHIP FLIGHT

Since the early days of civilization there have been many records of man's colourful and disastrous attempts to fly. The main trouble was that the bird made it all look so easy. It was pure envy that forced men to build wings to try to imitate her but man's muscles were too weak to carry his weight and his home-made wings too small to have any chance of copying her.

## *The Earliest Attempts*

As early as 1060 we hear of a monk leaping from a high tower in Malmesbury, flapping his arms with desperate enthusiasm. One of the lucky ones, he escaped with only a broken leg. An Arab in Constantinople tried the same trick a century later and killed himself. There are other accounts of would-be aeronauts through the ages, all of

1

'As early as 1060 a monk leapt from a high tower ...'

whom met with a similar lack of success. This was due mainly to man's conviction that imitation wings were the only path to success.

Even Leonardo da Vinci, the famous artist and scientist who died in 1519, left behind a detailed design for a flying machine operated by pulleys and ropes that flapped its wings like a bird. The actual model was never built but the drawings found after his death give us the first record of this interesting idea. Other people copied his designs and actually built wooden bird machines but there are no accounts of any leaving the ground. Before flying machines could be a success, man had to learn much more about the air itself, and how to harness it as a helpful force as he was already doing with the sails of ships and windmills. The first breakthrough came with the discovery of the hot air balloon.

## The Hot Air Balloon

It is a scientific law that hot air always rises. The Chinese had been aware of this fact since the fourteenth century and used to make small paper balloons which they filled with hot air and released on special feast days or celebrations. Sometimes the balloons caught fire whilst being filled but more often they soared away in the wind until the air cooled and brought them to earth again.

## The Montgolfiers

It was not until the eighteenth century, however, that the hot air balloon was really taken seriously, when two Frenchmen, the brothers Joseph and Etienne Montgolfier began to experiment with paper balloons filled with hot air. The brothers were not scientists but paper manufacturers and it seemed unlikely that they should suddenly take such an interest in flight. Legend has it that Joseph watched while his wife's petticoat, drying in front of the fire, filled with smoke and drifted up the chimney! He decided that it was smoke and not hot air that had the ability to push things upwards. Experimenting further with this theory, he came to the astonishing conclusion that the smoke produced by burning straw or chopped wood was lightest and produced the best results when lifting paper balloons.

He made his first small balloon in November 1782 from a spherical envelope of silk with an open neck at the base. Just by holding burning paper beneath the hole in the bag, he found that it rose to the ceiling of his room in a matter of seconds. When he performed this experiment again for his brother in the open air, Etienne was extremely impressed and began to work with Joseph to build bigger and better balloons.

In June 1783 they built a balloon of linen which measured 110 feet around with a 16-foot wooden base, and on June 5th gave a public demonstration attended by scores of disbelievers. Much to everyone's amazement the balloon took lazily to the air with its pot of glowing coals suspended underneath. After covering about one and a half miles in ten minutes the hot air cooled and it drifted peacefully to earth where it was set upon by terrified peasants who thought they were being invaded by monsters from outer space! By the time the Montgolfiers appeared at the scene it had been torn to shreds by pitchforks but as it had served its purpose they weren't really too concerned about its fate.

Ascent of the Montgolfier Balloon in 1783
*Courtesy Radio Times Hulton Picture Library*

The news of this first flight created quite a stir in the French capital, Paris, and the Montgolfiers were asked to show off their skill before King Louis and Queen Marie-Antoinette. They built another balloon elaborately decorated in blue and gold and bearing the royal insignia but whilst they were holding a rehearsal for the big day a sudden freak storm blew up and completely wrecked the new balloon. With only a week to go before they had to appear at the King's palace in Versailles, a new balloon had to be built. The Montgolfiers and their helpers worked around the clock to produce a new one, made of much stronger cotton and within four days it was ready.

Everyone had hoped that this time a man would actually fly with the balloon but there seemed to be a curious lack of volunteers. So, as a compromise, a duck, a cock and a sheep were chosen as passengers for the epic voyage. Suspended from the balloon in a wicker cage they rose to a height of over two miles and eight minutes later descended into the forest of Vaucresson two miles away.

When the observers reached the scene the basket had been broken open but the sheep was grazing contentedly and the duck also seemed to be no worse for his adventures. There was some alarm, however, when it was discovered that the cock had a damaged wing, obviously a result of its perilous journey. Everyone's fears were put to rest when many witnesses agreed that they had seen the sheep kick the cock before the flight had even started! So now no barrier stood in the way of the first aeronauts.

'The sheep was grazing contentedly and the duck also seemed to be no worse for his adventures . . .'

## Manned Ascent

The first of these was to have been a condemned prisoner but a young scientist, Pilâtre de Rozier, took pity on the prisoner's obvious terror and took his place. On Wednesday, October 15th, a new Montgolfier balloon was inflated ready to carry its first human passenger.

7

The balloon was about 75 feet high with a diameter of 49 feet. Its neck was 16 feet across, and a wrought-iron fire basket was hung underneath it.

Round the fire basket the brothers constructed a wickerwork gallery to carry the balloon's solitary passenger. During the experiment, he was to feed the brazier with straw, and keep the balloon in the sky as long as possible. As a safety precaution the balloon was secured to the ground by 84 feet of rope and de Rozier managed to stay on the end of it for 4 minutes 25 seconds before allowing it to sink gently to the ground. Delighted by his success, he made several more of these captive flights, gradually attempting higher altitudes and staying aloft for longer periods, until he finally reached a record height of 324 feet and managed to stay there for 9 minutes.

## Free Flight

The next step was to sever the cord that tied the captive balloon to its earth mother and let it take its chance with the elements. This was not as easy as it sounds. At that time there was widespread suspicion about the terrors that lay in wait for people who ventured into the upper air, a similar suspicion to that we felt about men venturing into outer space only a few years ago. The King

Ascent of the Montgolfier Balloon before the French Royal Family

*Courtesy Radio Times Hulton Picture Library*

himself was very doubtful about the wisdom of the plan but, in the interests of science, he allowed a demonstration to be fixed for November 20th in the Bois de Boulogne, Paris.

A large crowd gathered on the day of the event but a rainstorm blew up and the flight had to be cancelled. On the following day the skies were still overcast and threatening but the balloon was inflated and held in readiness. The Montgolfiers, still anxious about weather conditions, insisted that de Rozier should make one last trial run with the balloon tethered to the ground. De Rozier agreed to this but almost as soon as the balloon left the ground it was seized by a savage wind which badly ripped the fabric and almost wrecked it. For the crowd, who had been patiently waiting for some time, this was the last straw. Their mood of eager expectancy turned to one of menace.

The Montgolfiers, remembering the fate of a previous balloon that was torn to pieces by a hostile mob, quickly recruited a team of women to sew up the rents in the fabric and two hours later it was ready for inflation once more. At 1.45 the balloon bearing de Rozier and his friend the Marquis d'Arlandes rose from the platform and, like a lovely fragile bubble, drifted gently towards the River Seine. After the tension of the last few hours this triumphant moment was pure joy for the aeronauts and they laughed and waved their hats at the upturned, open-mouthed faces below. Unfortunately, due to the delicate balance of the balloon, they had to stand one at either side of its gallery and were completely unable to see each other!

In spite of this they took turns at stoking the fire with straw and gazing in astonishment as the familiar map of Paris unfolded under their eyes; the famous landmarks, the various bends in the River Seine – to these first aeronauts they were a wonderful sight indeed. Their attention was quickly drawn back to the balloon, however, when they realized that the fire had already caught hold of its delicate fabric and holes were beginning to appear on its underside. With a pail of water and sponges they worked hard to put out the flames before the whole fabric of the envelope caught alight.

They managed to land twenty-five minutes later completely unharmed after crawling out from under the collapsed balloon which had fallen over them. The world's first balloon flight was a success!

## The Hydrogen Balloon

During the excitement caused by the Montgolfiers and their hot air balloons a young physicist, Jacques Charles, had been experimenting with a lighter-than-air gas called hydrogen. Aided by two craftsmen brothers Ainé and Cadet Robert he made a balloon of silk coated with a solution of rubber strong enough, he hoped, to contain this new highly inflammable gas. The first hydrogen balloon was fairly successful on its maiden flight although it was a more difficult process to fill the envelope of the balloon with hydrogen rather than hot air.

After a short flight it was thought to have exploded but as once again frightened villagers tore it to pieces it was difficult to tell exactly why or how it had come to grief.

A second hydrogen balloon was then built which turned out to be a great advance on anything previously constructed. Once more the envelope was made of rubberized silk, this time in red and yellow sections and the perfectly round balloon measured 27 feet 6 inches across. Charles had designed a new self-operating valve which as well as allowing the balloon to be filled with gas also allowed it to escape freely when it expanded due to the heat of the sun. In this way there was much less risk of the balloon bursting. On top of the balloon he had built another flap valve which could be operated by the aeronaut pulling a cord whenever he wanted to descend again. Sacks of sand or water were to be carried as ballast which could be jettisoned to lighten the balloon.

The top half of the balloon was criss-crossed with a webbing of rope from which was carefully suspended a boat-shaped wickerwork basket made in a very elaborate gilded design. This impressive sight brought literally half the population of Paris, about 400,000 people, into the gardens of the Tuileries on a radiant first day of December in 1783.

By contrast to the slow and lumbering ascents of the hot air balloons, the frisky striped hydrogen balloon leapt from the ground like a cork from a bottle. The

Jacques Charles descending after making the first aerial voyage in a hydrogen balloon

*Courtesy Radio Times Hulton Picture Library*

crowd roared its approval and the aeronauts feeling completely lightheaded, soared over the rooftops of Paris, crossing and re-crossing the Seine, drinking champagne and even exchanging happy shouts with well-wishers below. After travelling twenty-seven miles, they managed a perfect landing by careful use of the gas valves.

This was the last flight of 1783 and it was astonishing to think that it was only a year since Montgolfier had begun his first simple experiments. Balloons were now the rage of Paris and no one could talk or think of anything else. Everyone was making and flying miniature balloons in elaborate shapes and sizes all over the sky. The craze got so out of hand that the Government decided that toy balloons were a serious fire risk and banned them! Full-sized balloons, both hydrogen and hot air, were being inflated like mushrooms both in France and abroad. Amongst the dashing young men of the times a fierce rivalry existed, particularly when they realized how much money could be made by giving spectacular exhibitions of their skill.

### The Fearless Blanchards

Women were not slow to jump on the bandwagon either; one of them, Marie Blanchard, wife of a French balloonist, spent many hours in the air sleeping, she claimed, in her tiny wickerwork car. Her unusual sleeping habits ended disastrously on one flight in 1819 when the hydrogen, escaping from the neck of the balloon, caught alight and quickly consumed the bag of gas. The basket dropped like a stone, tipping Madame Blanchard out on the way. After tumbling helplessly down the roof of a house, she died instantly on the pavement below.

On June 7th, 1785, fourteen months after man's first ascent, Pierre Blanchard, husband of the ill-fated Marie (who had not yet met with her fatal accident), decided to cross the Channel in a balloon. He took with him on this reckless gamble an American doctor, Judge Jeffries, for the sole reason that he could afford to pay the expenses and Blanchard certainly needed the money and, of course, the publicity! The balloon itself created a great deal of interest at the time, due to its newly designed basket. It was shaped like a bathtub with the addition of four rudders and a rear fin with which Blanchard proposed to steer the balloon through the sky like a boat.

The intrepid balloonists flew off from Dover dressed in their Sunday best and in lighthearted mood. They wanted to look smart for their triumphal return to France. However, within minutes of their ascent they realized that the wind had changed and they were being pulled in the wrong direction. This seemed the ideal opportunity for trying out the new rudders but much to Blanchard's disgust no amount of pulling on them made the slightest difference. Fortunately the wind changed of its own accord and they proceeded slowly across the endless stretch of grey water towards France, the

balloon all the time sinking closer and closer to the waves. Blanchard and Jeffries hurled out sack after sack of ballast, the books and instruments that they had brought, food and champagne and finally the rudders themselves. As the French coast came into view the sea was still much too close for comfort!

Blanchard, determined not to be defeated when they were so near land, began to strip off his clothes and threw them into the water, too. Jeffries reluctantly followed suit and may have just tipped the balance as another breeze suddenly lifted them high in the air, over the coast of France, and dropped them in the middle of a forest. Shaking with cold and relief the half-naked men leaned out of the basket and grabbed at the branches of a tree. Jeffries eventually caught and held one whilst Blanchard released the hydrogen from the envelope. They tumbled out of the basket onto the ground and with their teeth chattering, crept miserably from the forest to look for help. Certainly not the triumphant entry that they had in mind, but this did not detract at all from their amazing achievement! On the one hand they had learned that man cannot steer through the air with rudders as he can through water and on the other they had proved that even with a little knowledge man could already travel long distances by air.

Madame Blanchard falls to her death
*Courtesy Radio Times Hulton Picture Library*

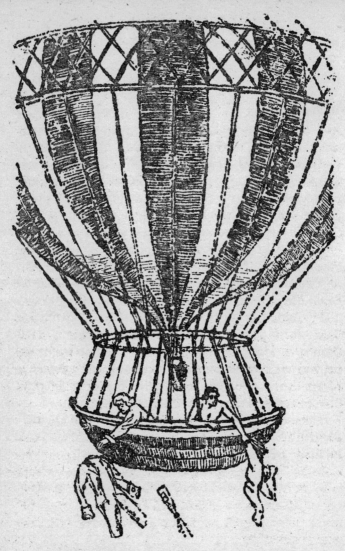

'Blanchard began to strip off his clothes and threw them into the water ...'

## Balloons for War

In the years that followed balloons were used extensively in many countries for military and commercial purposes. In 1794 the French Aerostatic Corps was formed, the world's first air force! A fleet of captive balloons was used as an observation post, and it was claimed that when the balloons ascended to the limit of their cable the operator could clearly observe a radius of eighteen miles. The Austrians, their enemy at that time, thought the appearance of the spy balloon over the horizon highly unchivalrous and proceeded to shoot at them. This turned the Balloon Corps into heroes overnight. Their popularity with the ladies and the attention they received became so embarrassing that it contributed to the unit's being disbanded. However, it led the way to the acceptance of the balloon for observation purposes and even as late as World War II in 1939 observation balloons were still being used.

In the years that followed many new methods were tried for steering balloons. Some had a hint of the future in them but most were unsuccessful and often ridiculous. There were plans for sails, screws and paddles to be worked by hand, clockwork or possibly airborne horses! There was a serious school of thought that believed trained eagles harnessed to the balloon to be the only possible answer!

In 1834 the European Aeronautical Society was

formed. In spite of its grand title it was really rather a dubious little company claiming to provide a regular airship service between London and Paris. The first airship of this fleet was designed to be rowed through the sky by vast oars and it was called *The Eagle*! The first airline office opened in London's Soho to await the *Eagle*'s arrival from France. However, while being inflated for the trip the unlucky balloon shot into the air and exploded. The crowd moved in and destroyed the remaining fragments in the traditional manner. In those days credit was never given for effort alone!

A second ship was built in London but when the company was found unable to pay its creditors the Sheriff of Middlesex carted the balloon away in a wagon and the company went bankrupt!

## *The Coming of the Dirigible or Powered Airship*

In 1852 a Frenchman, Henri Giffard, built a streamlined balloon 144 feet long and pointed at each end. It was powered by a tiny steam engine that turned a propeller. Giffard, looking very grand in a top hat and frock coat, stepped into his creation and proved, mainly by going round in small circles, that his craft could manoeuvre in any direction he chose. The only snag was that if there was any slight wind or other adverse weather condition it would become as out of control as any free balloon. In spite of this it did seem as if the world was on the verge of an exciting breakthrough. He called it an airship but it was another thirty-two years before a stronger airship was built.

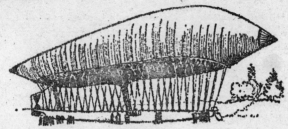

*La France*

This was *La France*, an airship powered by an electric motor. It managed to travel as fast as twelve miles an hour but unfortunately its batteries wore out far too quickly. It was now obvious that unless a really efficient power unit was invented there was only a limited future for the steerable balloon.

Eventually a Brazilian, Santos-Dumont, built a series of airships with the recently invented petrol engine as their motive power. To begin with they were unpredictable and rather dangerous but Dumont was not only a very wealthy man, he was also very determined. As the series progressed, however, his tiny airships became surprisingly reliable and suddenly he was the most talked-about man in Europe. Although he contributed little to the design of the airship itself, he must be given credit for being the inventor of the world's first practical airship.

## Germany Enters the Field

Whilst Santos-Dumont had the world at his feet another promising airship designer was quietly planning and scheming in Germany. His name was Count Ferdinand von Zeppelin and he was to become the father of the rigid airship as we know it today. In his opinion the inflatable envelope of the present airship was quite unsafe – just one small leak in this bag could bring about complete disaster. He proposed to build an airship of many independent gas cells, contained inside a rigid framework of metal girders and encased in a linen cover. He built his first prototype in 1902 and never looked back. He went on to build airship after airship, proving without doubt that his 'rigid' model was far ahead of any other lighter-than-air craft that was being built at the time. Despite various setbacks Zeppelin never lost his obsession for airships and built many of them for both peaceful and warlike purposes.

## The Airship Goes to War

The airships Zeppelin built for the German Government became deadly weapons in World War I when

Count Ferdinand von Zeppelin

they mounted a series of bomb attacks on England. The bombs were often miles off target but these first air-raids terrified and demoralized the British people until they were able to build aircraft and guns advanced enough to deal with the Zeppelins. The Germans were, at that time, the only country using rigid airships. Italy, France and Great Britain were still using the non-rigid inflatable observation balloons. They were shaped like miniature Zeppelins and were nicknamed 'blimps', apparently because of their 'limp' envelopes. They performed a useful service during the war, particularly when escorting convoys of merchant ships. From their great height they were able to spot submarines, another new menace,* and drop bombs on them.

1918 brought the end of the war and Germany's defeat. After the surrender there were seven military airships left in working order and these were divided between Great Britain, France, Italy and Japan. Although few of them were ever flown again they were used as prototypes for new airship projects.

## The Race to Cross the Atlantic

It was an exciting time now for aviators as, apart from the mighty airship, the tiny aeroplane was gaining in

* You can read how submarines were invented in *Submarines!* another Piccolo Book by the same author.

popularity and there was great rivalry between the supporters of the two types of aircraft as to which should be the first to cross the Atlantic. In June 1919 Captain Alcock and Lieutenant Arthur Brown in a Vickers bomber struggled across in very poor weather from Newfoundland to Ireland in sixteen hours.

The British then decided to try the newest of their airship fleet, the R34, flying the opposite way against fierce Atlantic winds. This time it took four days before the R34 reached Long Island, which in itself was breaking the airship endurance record. Then she turned around and flew back again breaking several more records.

This particular flight started a boom in the airship business. Now everyone wanted one. Airship building began in earnest and each new ship claimed to be bigger and better than the last. The British were asked to build one for the USA. This model, the R38, was 695 feet long and 85 feet across. Apart from her ability to lift 83 tons she was also a thing of great beauty. Appearances can be deceptive, however, and the crew found her very hard to handle. After her third hair-raising flight, when she got badly out of control and almost crashed, she was grounded and completely overhauled. On her next test run, the final one before being handed over to the United States, she completely broke in half over Hull and deposited her crew into the river Humber. Only 5 of the 49 survived. It was the sort of disaster that was to happen only too often in the subsequent history of airships.

The L72 *or Dixmude*

### *The Dixmude Mystery*

In 1923 one of the long-range German airships, the L72, claimed by France as spoils of war, was re-christened *Dixmude* and flown in a record-breaking $4\frac{1}{2}$ thousand-mile tour of North Africa lasting 118 hours and 41 minutes. She set off again on December 18th with a crew of fifty to try to break her own record. After being sighted flying over Tunis she disappeared without trace and extensive searches on land and sea revealed nothing. A few days after Christmas a Sicilian fisher-

26

man found the body of the ship's captain in one of his nets. This confirmed the opinion that the *Dixmude* had met a tragic end yet no further sign of the wreckage or the other 49 men has ever been found, and the reason for her sudden disappearance is shrouded in mystery.

Surprisingly this did nothing to stem the tide of enthusiasm of airship builders; if anything it gave the airship a romantic and mysterious image and the silver giants continued to grow and multiply. In 1926 three countries, America, Norway and Italy combined in an attempt to fly an airship the *Norge*, over the North Pole. In spite of severe icing-up and dense Arctic fog the airship reached the North Pole safely and circled it gracefully whilst the flags of the three nations were dropped onto the snow.

## Britain Gives Up

In 1929 a new experimental airship programme was launched in Britain that was to cost £1,350,000 and included the building of two rigid airships, the R100 and her sister ship the R101. These were large enough to carry heavy loads and were equipped to travel across the world. The R100 had some teething troubles but proved successful in her trials. The R101, on the other hand, was always in trouble, her major problem being overweight. On her maiden flight to India she crashed in Northern France, killing forty-eight people.

At this the British felt they had now had enough of airships. The successful R100 was broken up and sold as scrap and the whole airship project was abandoned. There had been other similar airship accidents and it was becoming obvious that so long as inflammable hydrogen was used any minor accident could turn into a full-scale funeral pyre.

The R100

## *America Carries On*

America had experienced similar tragic accidents with her airships and after the crash of one of her greatest,

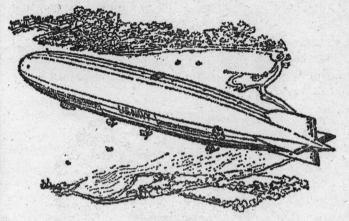

The *Shenandoah*

The *Shenandoah*, they almost gave up too. At this time however, they had the monopoly of a new non-inflammable gas called helium and the American military argued that if helium proved successful they could have an airship fleet to lead the world. On the strength of this theory two new super-airships were born.

The *Akron*, built in 1931, was a flying wonder 785 feet long with a lifting power of over 200 tons. The use of helium now allowed her eight powerful engines to be housed inside the main bulk of the airship, making it no longer necessary for crewmen to climb up and down precarious ladders in all weathers. Only the propellers projected from the body of the ship giving her a more streamlined and less air-resistant shape. Amidships she had a hangar equipped to receive *five* fighter planes and in the belly of the ship an electrically-powered trapeze with which to lift and lower them. If she could fly successfully, her possibilities were endless!

Extensive trials found her graceful in flight, efficient in handling the comings and goings of her visiting aircraft and very useful in long-range reconnaissance missions. In fact, everything the Americans had ever hoped for. All over the world the *Akron*'s progress was watched with envy and fascination; the largest, most sophisticated airship ever flown, she was a credit to her creators. It was a blow for America when, on a stormy night in April 1933, she was broken open and tossed into the sea. Most of her crew managed to jump out but as they struggled to keep afloat the wounded airship rolled back and forth over them like a gigantic rolling-pin, forcing them down under the waves. Out of the seventy-six men on board, only three survived. No one had thought of including life-jackets amongst the impressive equipment she carried.

Just a month before the *Akron* tragedy her sister-ship the *Macon* was brought from her hangar. Almost a carbon copy of the *Akron*, she followed firmly in her sister-ship's footsteps. She handled perfectly during trials and for two years took part in many fleet manoeuvres and exercises. Once more the American people pushed the previous disasters to the back of their minds and accepted the *Macon* as being totally reliable.

Yet her fate was ironically similar to that of the

*Akron*. In another storm over the sea the *Macon*'s tail fin broke off and she began to sink slowly towards the water. This time, happily, the crew had time to release their rafts and put on the life-jackets which were now carried on board. Of the eighty-three crew members only two died. The low death toll did nothing to appease the anger of the American people. The end of the *Macon* caused a tremendous uproar in the press and convinced everyone that lighter-than-air ships were totally unsafe. The *Macon* was the last rigid airship to be built in America.

## The Last Tragedy

By this time nearly all the major nations had experimented with airships, and nearly all had, after some disastrous accident, abandoned their airship projects for the foreseeable future. Throughout the string of airship tragedies that had shocked the world only Germany stood firm in its utter belief in her Zeppelins. The famous *Graf Zeppelin* had flown millions of miles safely in passenger service and there was always a long waiting list for places on board. In fact, apart from the wartime Zeppelins, not a single death had occurred in the history of the German lighter-than-air ships.

In 1936 a new commercial Zeppelin, the *Hindenburg* was built, even larger than the *Graf Zeppelin* and designed for use with the new helium gas. However, by the time it was launched America had refused to sell any of her helium gas to Germany and inflammable hydrogen had to be used instead. She seemed as safe and reliable as the *Graf Zeppelin* but in 1937 she burst into flames and was consumed within minutes. After this final catastrophe not even Germany would have dared to fly another airship. Even the *Graf Zeppelin*, after years of faithful service, had to be deflated. For the vast, rigid airships it was the end of an era. Yet these romantic giants will never be forgotten for they taught the world a useful lesson and there are still many men who dream about their return.

The *Europa*

Yet 1972 witnessed the birth of a new airship, the *Europa*, at Cardington, England. Built by the Goodyear Tyre and Rubber Company, she is powered by twin Continental 6-cylinder engines, each of 210 hp, which

The *Hindenburg* ablaze

gives her a maximum speed of 50 mph, a cruising speed of 35 mph and a 500-mile range.

The *Europa* has a 23-feet-long cab which can carry the pilot and up to 6 passengers. Unlike the rigid airships of the 1930s she has no metal framework to maintain her shape. Her envelope, made of Neoprene-coated Dacron, is pumped full of lifting gas and also contains cells of air to compensate for the differences of pressure. An unusual feature of this new airship is the 7000 lights built into her sides to flash after-dark messages and advertisements.

The *Europa* was launched in March 1972 and three days later made her first flight which was very successful and lasted three hours. Plans were made for her to carry TV cameras, in order to cover major sporting events, including the Derby and the Cup Final. Unfortunately on April 20th, 1972, the airship came adrift and blew into the grounds of a farmhouse in the village of Cotton End, where it sustained £400,000 worth of damage.

It was soon repaired, however, and crossed the Channel for the first time in July. Since then it has been used to televise a number of motor races, including the British Grand Prix, and also covered the yachting events in the Olympic Games held in Kiel.

*Europa*, we will watch your progress with interest!

## Chapter Two

## THE MAN WHO REFUSED
## TO BE BEATEN

In 1873 a male child was born to a family of Brazilians named Santos-Dumont. The boy was given the name Alberto and, the apple of his father's eye, was given the many advantages of a wealthy upbringing. His father owned the thriving coffee-plantation where Alberto spent his childhood and it proved to be an exciting place for a boy who was interested in machines and inventions. At the age of seven he was allowed to drive a 'locomobile', a steam traction engine with broad wheels that was used in the fields.

When he was twelve he was driving a locomotive engine that hauled train-loads of green coffee beans over the sixty miles of the plantation railway and when his father and brothers were out of the way he rushed down to the coffee factory where he spent hours playing with the coffee-processing machines. There is a great deal of varied and complicated machinery to be found in a coffee factory and it was not long before Alberto could repair or replace any part of it.

Besides being very clever with his hands, Alberto was

also a dreamer and spent his spare time reading the popular adventure stories of the day and making his own plans and inventions. Sometimes he made aeroplanes from pieces of straw and sometimes he made balloons. Every year during the Brazilian festival days he made fleets of tiny Montgolfiers and he and his family watched with delight when they drifted up and away.

Alberto saw his first real balloon at a fair when he was fifteen. He had already read about Montgolfier and Pilâtre de Rozier with interest and he now began to dream of building his own airships and flying machines. However, he was forced to keep such ideas to himself. To talk of such a thing in the Brazil of the 1880s would have indicated complete insanity. The people who actually flew balloons were put in the same category as actors or acrobats, quite interesting and amusing, but certainly not the sort of people a respectable planter's son would have anything to do with!

In 1891 the Santos-Dumont family decided to take a holiday in Paris. Naturally Alberto was delighted because it was now forty years since the Frenchman Giffard had flown his steerable airship, and Alberto was looking forward to seeing many improved versions of it flying all over France. He was amazed to find on his arrival that there were still only spherical balloons in use. No one had improved on Giffard's 'airship' and for many years no cigar-shaped balloons had been seen in the skies of Europe. However, still filled with enthusiasm, Alberto found out the address of a professional balloonist and went to his house to try and hire him for an ascent.

To begin with, the aeronaut thought Dumont, now eighteen, far too young but eventually agreed to take him up, provided Alberto paid him 1,200 dollars, plus the railway fares for himself and the balloon from wherever they landed back into Paris. Alberto, wealthy but not foolish, had second thoughts; after all twelve hundred dollars was an awful lot of money. So he gave up the idea of ballooning and became interested in the automobile instead.

He took another holiday in Paris a year later and this time found a balloonist who was willing to take him up for 250 dollars, so this time he decided to try. The great day dawned, mild and fresh: Alberto arrived at the starting point early so as not to miss anything. The limp and lifeless balloon was spread over the grass waiting to be brought to life. A team of workmen began to pump the hydrogen into it and it grew larger and more powerful by the second! By 11 AM the balloon was attached to its wicker basket and straining to be off. Alberto and the balloonist, M. Machuron stood in the wicker basket holding up the ballast. 'Let go all,' M. Machuron shouted and immediately they floated away from the ground.

They left the earth calmly as if coaxed by a gentle wind, yet as they looked down the fields and trees seemed to be dropping away from them at a fantastic rate! Alberto was spellbound by the sight and glorious feeling of being whisked through the skies in this particular manner and, after what seemed like only minutes, they heard a village church ringing out the noonday bells. Santos-Dumont, who believed in doing things in style, had brought with him a lunch hamper. This delicious spread included hard-boiled eggs, cold roast beef and chicken, cheese, ice-cream, fruit, champagne, coffee and liqueurs! He swore it was the most enjoyable meal he had ever had.

Soon after lunch a forceful wind began to pull and tug at the balloon and, as they were running short of ballast anyway, they decided to bring the balloon down. They landed safely, and after letting the gas escape they folded up the spent balloon and put it inside the wicker basket. They hired someone to transport them to the nearest station and, after heaving the 440-pound basket on board, they travelled the sixty miles back to Paris.

Alberto liked ballooning so much that he decided he would have a balloon built for his own use and commissioned a M. Lachambre to build it for him. It was to be a very small balloon made of the lightest and toughest Japanese silk and when it was finished he called it the *Brazil*. The *Brazil* turned out to be a great success, particularly as its size made it so easy to transport between one trip and the next. Alberto had many happy adventures in it and soon made a name for himself as a very able balloonist. Yet Santos-Dumont was not really satisfied; he still thought a lot about Giffard's dirigible (or steerable) balloon built many years before and soon convinced himself that he could design a dirigible balloon that would work even better.

He wanted to try the new petrol engine which was rapidly proving itself to be a reliable motive power and decided to build an elongated balloon light enough to carry a tiny $3\frac{1}{2}$ horse-power tricycle motor which weighed only 66 pounds. At first M. Lachambre, the balloon-maker, flatly refused to take part in what he considered to be utter recklessness but after Alberto had threatened to sew the balloon with his own hands, he gave in and made the balloon he wanted in Japanese silk.

In September 1898 Dumont was ready to take his brainchild out into the open air. His friends, most of whom were aeronauts, had heard that he intended to carry a petrol engine, which often gave off red-hot sparks, under a balloon filled with the highly inflammable hydrogen gas. They were absolutely horrified, and begged him to use the far safer electric engine.

Alberto would not listen to them and on September 20th *Santos-Dumont* No 1 was brought out for her first flight. Curious people from all walks of life had turned out to witness this amazing spectacle and all were horrified when the flight began with a series of loud explosions from the bad-tempered petrol engine. In spite of this unpromising start, No 1 eventually sailed off confidently into the wind with Dumont erect at the helm properly attired in bowler hat and high stiff collar. The airship had no trouble reaching the end of the field and the crowd held its breath. Would she be able to turn round? With surprising grace the fragile craft did swing round and, gathering speed, soared over the heads of the cheering crowd. Not content with this, the daring little Brazilian went on to demonstrate figure-of-eight turns and other unheard-of feats, finishing with a bumpy, but safe, landing.

Santos-Dumont was a very small man with a large black moustache who always prided himself on being fashionably and expensively dressed. Now, almost over-night, he became the dashing, impetuous Alberto, the darling of Paris, admired by men and adored by women. To begin with he was startled by his popularity but he soon settled down to enjoy it! However, he was too level-headed to become diverted from his main interest in life, the dirigible, and in the years that followed he went on to build fourteen of them. At the start he had trouble getting the elongated balloons to stay rigid and whilst in flight his No 2 balloon bent completely in half so that both ends were pointing to the sky. He emerged unscathed!

Gradually he modified the shape of his balloon so that, although still streamlined, they were thicker in the middle and consequently safer. He dispensed with the wicker basket and built a framework of tightly-stretched cords connected to a bamboo pole. In the middle of this spider's web was perched a bicycle seat on which Dumont sat and from which he controlled the ballast, engine and steering at the same time as keeping his balance. Not an easy task! His tiny airships became a familiar sight in the streets of Paris where, whenever the mood took him, he would drop down to his favourite restaurant, throw out his mooring-line and stay for lunch. On at least one occasion he was known to have tethered his airship to the lamp-post by his own front door and gone in for a meal or a drink.

In 1900 whilst Alberto was flying his No 4, a member of the Paris Aèro Club, Henri Deutsch de la Meurthe, offered a prize of 125,000 francs (a very large sum worth about £40,000 today) to the first dirigible balloon that could fly from the Aèro Club at St Cloud, round the Eiffel Tower and back all in the space of half an hour. This was just the sort of challenge that the dynamic Brazilian thrived on although he, of course, had no need of the money. With this exciting prize in view he began to design and construct his fifth airship. No 5 was bigger than his previous creations and had a triangular wooden keel which was slung several feet under the gas-filled envelope. This time he mounted a wicker basket on it instead of the vulnerable bicycle seat and there was still plenty of room to take the heavy 12-horsepower engine. He also had a heavy rope hanging down onto the platform. By carefully swinging the rope, which altered the ship's centre of gravity, he was able to tip the nose of the balloon up or down according to his wishes.

He gave it a test flight on July 12th and everything worked extremely well, so he sent notice to the Paris Aèro Club to say that he would be making his attempt the following day. The news of Alberto's bid for the prize spread through Paris like wildfire and by dawn on July 13th a seething crowd had gathered outside the Club. The judges arrived and included Prince Roland Bonaparte and Deutsch de la Meurthe, the man who was giving the prize.

At 6.41 Alberto cast off and with the ease of a bird reached the Eiffel Tower and rounded it in only ten minutes. After this his troubles really started! On the way back an unexpectedly fierce head-wind slowed him down to a crawl. In spite of it he managed to battle on

'With the ease of a bird reached the Eiffel Tower and
rounded it in only ten minutes . . .'

until he came within shouting distance of the time-keepers at St Cloud. At that moment, completely without warning, the engine stopped dead. The airship, robbed of its power, was whisked off by the wind and thrown into the tallest chestnut tree in a park belonging to M. Edmond de Rothschild.

His sudden unexpected arrival brought all the Rothschild family and their servants out of their villa within seconds. They were amazed to find the airship still in one piece and Dumont completely unharmed! Although the balloon's envelope was now empty of gas the only damage Alberto could see was a few tears in the fabric. In spite of his failure to win the prize he had been very lucky to land so safely!

This accident happened very near to another large house where a Brazilian princess, Isabel, Comtesse d'Eu lived. When she heard of his misfortune, and that he would be a long time freeing the airship from the tree, she sent lunch up into the tree for him. Afterwards she invited him to come and tell her about his adventures and a few days later she sent him this letter:

August 1st, 1901

Monsieur Santos-Dumont,

Here is a medal of St Benedict that protects against accidents. Accept it, and wear it at your watch-chain, in your card-case or around your neck. I send it to you thinking of your good mother and praying God to help you always and to make you work for the glory of our country.

(Signed) Isabel, Comtesse d'Eu

Alberto was delighted with the medal and had it put onto a thin gold chain. So ended a happy interlude but it brought him no nearer his goal – the Deutsch prize!

The airship was duly repaired and Alberto took her for several successful flights over the safe green grass of the Longchamps racecourse. He was delighted with her speed and apparent reliability and once again applied to enter for the Deutsch prize. On August 8th, 1901, at 6.30 AM and in the presence of the Aèro Club judges, Santos-Dumont started again for the Eiffel Tower. As before, he set off very smoothly and in the very fast time of nine minutes he had turned around the tower and started the return trip to St Cloud. This time the weather was much more favourable but Alberto realized that he was rapidly losing gas from a faulty automatic valve. He knew that the safest course of action would be to land and repair it straight away but his pride would not stand for that and so, like the captain of a sinking ship, he stood hopefully at the rudder whilst the balloon gradually shrivelled up like a prune!

Then gradually the suspension wires began to sag. One of them caught inside the screw propeller as it revolved so Dumont had no choice but to stop the engine. As in his previous attempt the balloon was caught by a powerful wind and forced back towards the tower it had come from. At the same time because of the loss of gas, it began to fall. Alberto was tempted to throw out some of his ballast to check his headlong fall but by then he was so near to the tower that he was terrified of being dashed to pieces against it. He watched helplessly as No 5 began to be swept down towards a street of hotels. This time there was no comfortable tree to fall in and the bag of the No 5 smashed with a loud explosion on top of two six-storey hotels. Once again Dumont was extremely lucky for the pole on which he was riding came to rest with one end on each building. Like a trapeze artist Alberto was left dangling somewhere in the middle, and he had to stay there until firemen came to rescue him!

After an uncomfortably long time they let a rope down from the roof to the little Brazilian and hauled him up looking just as spotless and perky as when he set out. The airship had not been so lucky. It had to be rescued from its perch in bits and pieces. Dumont's pride really took a knock this time and what had previously been just an interest in the Deutsch prize now became an obsession. That evening he sat down and designed *Santos-Dumont* No 6. It took twenty-two days of hard work before it was finished and inflated. This new airship was very similar in design to No 5 but Alberto had spent a lot of time on the details that had previously let him down, particularly the automatic air valve which last time had proved faulty.

On October 18th, after weeks of trials, he sent the necessary telegrams to the Aèro Club judges inviting them to be present for his next attempt which was to take place at 2 PM the next day. Alberto's friends considered him to be very rash as the weather that day was dreadful but he would not be put off, and even when the weather turned out to be just as unfavourable on the 19th he refused to cancel his bid for the prize. By 2 PM conditions had improved slightly but not enough to tempt most of the judges out of doors and only 5 of the

25 turned up to watch the proceedings. It was 2.42 before Alberto was ready to be off, and as soon as he left the ground a strong wind caught him sideways on. The No 6 was made of stern stuff and managed to hold a straight course in spite of it.

As in his previous flights Alberto reached the Tower in nine minutes. The return trip, he knew, would take a lot longer as he would be flying in the teeth of the same wind. What he did not expect was for the engine to stop working, which it did! It was an awful moment for Dumont as, in order to restart the engine, he would have to climb out of his basket and let go of the steering wheel. As he debated what to do the engine suddenly started again and he was able to continue. When No 6 was nearing the end of the course the same thing happened again, this time completely upsetting the balance of the airship and taking her up much higher in the air than Dumont intended.

By this time it was too late to bring the airship down before reaching the official timekeepers, so No 6 soared high over their heads like a horse who would not stop at the winning post. Dumont turned it around some way farther on and returned to the Aèro Club to find that he had passed over their heads 29 minutes and 31 seconds after starting out!

The crowd were overjoyed at what seemed like a certain victory but the judges were not in complete agreement. Deutsch de la Meurthe wanted to give the prize to Santos-Dumont but the other judges claimed that as he had not been able to land inside the 30-minute time limit he was disqualified! This began a violent argument between the judges but two weeks after the event Alberto was told officially that he had won the elusive prize. As he was already extremely rich he gave the money away. He gave 75,000 francs to the poor of Paris and the rest was distributed among the men who worked for him.

Alberto Santos-Dumont went on to build eight more airships, drawing on his own fortune to improve each one and he did much to popularize lighter-than-air aviation. In 1906 he built and flew his own aeroplane, being the first man in Europe to do so. However, his name will always be most respected and admired by those interested in dirigibles and balloons and he will always be remembered as the man who designed the world's first dirigible and actually made it fly successfully.

Chapter Three

## THE INCREDIBLE
## COUNT VON ZEPPELIN

At the same time that the French people were thrilling to the exploits of their idol, the dashing, fun-loving Santos-Dumont, the Germans were discovering that they too had a promising airship inventor in their midst, but this hero was cast in a different and more sinister mould! His name was Count Ferdinand von Zeppelin and having spent most of his life as a soldier he became interested in developing airships for more serious, war-like purposes.

He was born in 1838 on an island in Lake Constance, where Germany borders Switzerland. His mother was a French refugee and his father a wealthy German noble-man. It was the custom in those days for all German aristocrats to enter the army and when he was nineteen the young Count entered the Eighth Württemberg In-fantry regiment. Over the following years he led a very exciting life in the army, including a spell in America where he had the chance to take a trip in a balloon. It seems doubtful that he found ballooning very much to his taste, however, as afterwards he returned home to

Germany and settled down to become a soldier in the army of King Charles I of Württemberg and there is no record of any further balloon trips.

It was a busy time in Germany for the professional soldier and Zeppelin, now a captain, took part in many important battles and was finally awarded the Württemberg military cross for his bravery and resourcefulness. By 1887 he had been promoted to the rank of brigadier-general. The following year, however, a new Kaiser came to the throne who took a particular dislike to Ferdinand and for no apparent reason he was turned out of the army. He was now fifty-two years old.

It was a bitter blow for a man who had devoted his whole life to the service of his country and most men would have been broken and dispirited by it. Not so Count Zeppelin, a hard, determined man – in fact his personal battle was just beginning. He looked around methodically for a new field in which to make a name for himself and his attention was drawn to the accounts of Santos-Dumont and his dirigible balloons.

He became interested in the idea of building an airship himself with a possible view to providing a mail or cargo-carrying service. It was a new and exciting idea and once he had made up his mind to do it nothing could stand in his way. He and his family went to live in a house in Stuttgart where he found an engineer, Theodore Kober, who was willing to be his assistant. Zeppelin studied the work of other lighter-than-air designers but he became interested in a different theory. He felt that the non-rigid, inflatable type of airship favoured by most of the inventors of the day was no longer practical, especially in a ship built to carry a heavy cargo. It only needed one leak in any part of the envelope to bring the flight to a speedy end.

He planned to build an airship with a light metal frame of girders filled with individual balloons of air. He strongly believed that size rather than weight was the long-term answer for commercial airships. Most airshipmen thought the idea was outrageous!

When his plans were finished Zeppelin presented them to the German Government with a request for money to build a rigid airship. Not surprisingly he was turned down flat. This, of course, did not deter a man who had won a medal for resourcefulness, so he formed his own limited company and sold shares in it to the public. Before starting on the airship he built a shed for her and floated it on Lake Constance. Then his work on *Luftschiff* (Airship) *Zeppelin* 1 began in earnest.

She was to be 420 feet long, $38\frac{1}{2}$ feet across and powered by two Daimler 16 hp engines. She was finished in 1900 and prepared for her first test flight. On a warm afternoon in July a large throng gathered on the shores of Lake Constance on the German-Swiss border. As was the custom, when word leaked out about any mad invention, scores of people turned up to see it succeed or, more likely, fail disastrously. Today was no exception and after months of feverish activity in the vast sinister shed the crowd were more than a little dubious and frightened about what would eventually emerge. They were not disappointed! The great doors opened and the first thing they saw was a firebreathing steam tug whose fierce sparks, as she pulled the great bulk of the airship out, terrified the crowd before they even set eyes on the Zeppelin itself!

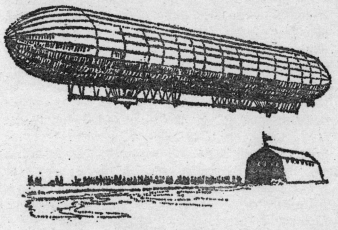

The first Zeppelin

The first rigid airship ever built, she was an historic sight well worth waiting for. Like an endless, angular sausage she lumbered gracelessly out of her hangar and came to rest on the placid surface of the lake. Two tiny gondolas were suspended beneath her great belly and between them hung a large, lead weight. This could be pulled forward or back to shift the ship's centre of gravity and raise or lower her great snout. In the first gondola rode the notorious Count, an impressive figure with his white hair and large, bristling moustache. At his command the unlikely vessel pulled herself reluctantly

from the surface of the lake looking about as far removed from a bird as it was possible to get. Yet fly she
did in spite of the jeers of the unbelievers!

Yet although her first flight was spectacular to behold
it was in reality a failure. Her progress through the air
was painfully slow and it was obvious to the observers
that Zeppelin and his crew seemed to be having great
difficulty in controlling her. The newspaper reporters
who had gathered for the occasion were very disappointed and criticized the experiment bitterly in the
papers the next morning.

Count Zeppelin was shaken. He was also angry that
the LZ1 had been given so little opportunity to prove
herself. He took her up on two other flights, each one
smoother than the last, but he was finally forced to agree
with the newspapers that much more time and money
was needed to produce an airship that was fast and yet
thoroughly reliable.

Yet what could he do? Already he owed money to
everyone, in fact he could not even pay his own employees. He had no choice but to sell up and dismantle his
great dirigible and its hangar. But Zeppelin seemed to
thrive on life's disappointments and it was not long
before he bounced back again with a scheme to hold a
lottery in the State of Württemberg to raise money for
his next airship.

The lottery was approved and Zeppelin went on to
build LZ2, almost an exact replica of LZ1, but with two
new powerful 85 hp engines as against the feeble 16 hp
ones he had used before. She had her first test in January
1906 – another spectacular failure! After making a

smooth, fast take-off her forward engine broke down and she began to float off rapidly to the North, completely out of control. There was consternation amongst the crew as they made frenzied but unsuccessful attempts to restart the obstinate engine. Zeppelin, of course, did not include the word panic in his own vocabulary and stood calmly at the helm instructing the men to gradually valve-off the hydrogen.

By doing this carefully they were able to bring the LZ2 safely to earth without doing any damage and Zeppelin was delighted to have saved his precious creation from destruction. He had the huge craft firmly lashed to the ground for the night, ready to transport her back to her hangar the following morning. Once again the Count's luck was out. A storm blew up during the night and for hours toyed with the captive airship, hurling her back and forth in her web of ropes before finally breaking her back.

Zeppelin was now laughed at openly, and the Kaiser, who had always disliked him was quoted as saying, 'Of all South Germans – the greatest donkey!' He had sold all his personal effects to help pay for the LZ2 and now he was truly destitute. Once more King William of Württemberg came to his aid and allowed him to hold another State Lottery.

The proceeds of this went to build the LZ3, an identical model to LZ2 but with much improved steering gear. Her maiden flight brought a perfection that amazed everyone except, of course, the obstinate old Count. The fact that in the space of a few hours he had changed from a crazy old fool into the most popular man in Germany did not seem to affect him in the least! Within weeks everyone was selling Zeppelin toys, picture postcards, brands of cigarettes and even Zeppelin winter coats and it became the fashionable thing to say that one had 'been to Lake Constance to see the airships!'

Not surprisingly the German Government had a quick change of heart and said that if Zeppelin could build an airship that could fly for 24 hours and land on the ground instead of on water they would agree to advance the money needed for the continuation of his experiments. So he built LZ4. She was 446 feet long, 42 feet across and fitted with 85 hp engines. She made a successful first flight on July 3rd followed by another flight eleven days later lasting twelve hours. Another historic flight on July 3rd took on board the King and Queen of Württemberg as passengers, making it the first royal airship flight ever! These record-breaking flights served merely to set the scene for the major 24-hour test which was to take place on August 4th, 1908.

Count von Zeppelin, now an imperious seventy-year old, took the controls as usual and at 6 AM the LZ4 began her long flight along the course of the River Rhine. The first part of the voyage was peaceful enough, and much enjoyed by the German people on the route who rushed out to watch the amazing ship pass overhead. After several hours, however, the forward engine developed a fault and the airship had to land for repairs. The engine was soon mended and the giant took off again. The flight continued smoothly until midnight, when once again the same engine began to overheat. Rather than land during the night Zeppelin slowed down the failing engine and nursed it along until morning. He then decided to land in a field near the Daimler automobile works in Stuttgart where the engines could be repaired.

After a perfect landing he strolled off to the nearest village and had breakfast at the local inn whilst the mechanics set to work on the repairs. Whilst he was enjoying his meal crowds began to gather round the airship and soldiers were called out to cordon off the area. Without any warning a fierce wind blew up and the hundred soldiers who were holding the Zeppelin down suddenly found they were wrestling with a demented beast. She twisted and rolled until finally the bow line snapped and the ship began to rise, taking some of the panic-stricken men with it.

There were yells of horror as men fell out of the sky, some from a great height, and the mighty beast surged away from them. She had travelled about half a mile when she exploded in a group of trees. For miles around the sky was lit with the flames of her pockets of hydrogen and Zeppelin seeing it, knew instantly what had happened. When he arrived at the scene he was horrified to find that several men had been burned and mutilated in their attempt to hold onto his creation. If nothing else had succeeded in bringing him to his knees, this tragedy did. At last, looking all of his seventy years, his proud spirit broken – he was ready to give up!

Yet it was the turning point in his career; in spite of the accident he became even more popular and from all over Germany little groups of people sent him small sums of money to continue his work. After a while the ageing count was encouraged to continue his search for the perfect airship and went back to flying the old LZ3 whilst he designed and built bigger ships for both military and commercial purposes. In 1901, in partnership with Dr Hugo Eckener, he built his first passenger airship, the *Deutschland*, which travelled at the hairraising speed of 45 mph and had a restaurant on board. Zeppelin and Eckener gradually set up dirigible airports all over Germany and eventually had a fleet of five airships including the *Deutschland* operating a regular passenger service.

There were still odd accidents but somehow this added to the glamour of airship travel and there was never any shortage of passengers. The *Deutschland* crashed in 1910 but all her passengers escaped safely. Another of the fleet, the *Schwaben*, was not so lucky and when she exploded in 1912 many people were injured and thirty-seven were killed.

Then came 1914 and war. This gave Zeppelin the chance to realize his ambition. Still a soldier at heart, he had always seen the dirigible as a powerful military weapon in war and it took him no time at all to convince the Government of the possibilities. Germany was the only country to possess military airships in 1914 and the tiny aeroplane was still unreliable, fragile and weaponless. The mighty airship seemed to be the ultimate threat and Germany, eager to press home her advantage, ordered the Zeppelin factory to turn out military airships as speedily as possible. The navy assigned thousands of men to the new Naval Airship Division and giant hangars were built in Germany and Belgium. The scene was now set for a cold-blooded aerial attack on defenceless Great Britain and her allies!

For the first part of the war the great silver monsters roamed the skies of Great Britain observing defences and troops and bombing cities. The mighty *Zeppelin* that had once awakened world-wide interest and admiration now became an object of dread. Yet as the war progressed the Allies gradually began to cut her down to size. They built up their defences and improved their anti-aircraft guns until they were powerful and deadly accurate. Most important of all, they improved their rickety airforce until it was now a force to be reckoned with. Together with the newly invented incendiary bullets fired by the pilots, it sent many a Zeppelin to an early grave.

When Germany lost the war in 1918 it was realized that the giant dirigibles, although terrifying to behold, were really far too bulky and too easy a target to be a success as offensive weapons. Despite this they had been of great value to Germany in the days before radar if only as long-range observers. They had fed back vital information particularly about the movements of the British and had earned even more respect for Count Zeppelin and his partner Hugo Eckener.

Doctor Eckener, now a skilled airship pilot and brilliant aeronautical designer, had been very close to Count Zeppelin in the last years of his life and when the old Count died at the age of seventy-nine in 1917 he was determined to carry on his work. After the war things were not too easy for the beaten Germans as they had to hand over their remaining airships to the victorious countries. However, Eckener, who had been responsible for designing and building most of the wartime aeroplanes, could at least keep his knowledge and experience to himself and fourteen years later was back at the top of the airship-building field.

A man totally dedicated to lighter-than-air craft, he still wanted to prove that commercial airships were practical for transcontinental and even round-the-world flights, and to prove it he built a super new passenger airship called *Graf Zeppelin*. She was a really vast ship 775 feet long and powered by five engines totalling 2,650 hp. She could carry 20 passengers and 12 tons of freight at a maximum of 60 miles an hour. The *Graf Zeppelin* (Graf is the German word for Count) was christened in von Zeppelin's honour by his daughter on July 8th, 1928, the 90th anniversary of his birth.

Her maiden flight almost ended in tragedy when she flew into a storm at full speed and badly ripped the fabric covering her port fin. In order to avert certain disaster the fabric had to be cut off whilst the ship was in flight and one of the men who volunteered to do it was Eckener's own son Knut. It was a hair-raising task for the four crewmen who clambered out on the wet girders. As they hacked at the loose fabric the storm lashed at them relentlessly, doing its utmost to hurl them into the sea below. In spite of this they completed the repair and the *Graf Zeppelin* was able to continue safely on its way.

After a flight of 6,168 miles and 111 hours it made a

triumphant landing in New Jersey, USA. From then on the *Graf Zeppelin*'s fabulous career really began. Before she retired ten years later she had made 590 flights, flown more than a million miles and carried 13,110 passengers without a single accident. Delighted with her success and the fame she had brought to Germany, Doctor Eckener built an even larger, more luxurious airship which he launched in 1936 and christened the *Hindenburg*.

This ship was 803 feet long by 146 feet across and her enormous Daimler-Benz diesel engines churned through the sky speeding along at over 80 miles an hour, making her the fastest ocean-going liner ever. Yet her performance was nothing compared with the lavishness of her décor. There were electrically-heated staterooms for the seventy passengers, each with hot and cold running water. The lounge and dining-room were exactly what one might expect to find in a luxury hotel, even including an aluminium grand piano! There were spacious promenade decks almost the length of the ship, with large picture windows. There was also an asbestos smoking room.

Doctor Eckener had only one worry about his new ship and that was the hydrogen that had been used to inflate her gas bags. He had originally built her to be filled with the new non-inflammable helium gas but America refused to sell him any in case it was used secretly in the building of military dirigibles. Adolf Hitler was now Chancellor of Germany and since he and his Nazi party had taken the country by storm, Germany had become hostile and aggressive towards her neighbours.

Many people feared that once again, the world was on the brink of war.

The *Hindenburg* proved to be as fast and reliable as her sister ship *Graf Zeppelin* and together the popular pair carried passengers on the Atlantic routes until the middle of 1937. One fine May day the *Hindenburg*, with a hundred on board, was preparing to land in New York. The flight, as was usual these days, had been completely smooth and routine. The New Yorkers were now so accustomed to seeing the Zeppelin that they hardly bothered to look up at her as she glided in, the afternoon sun picking out the sinister black swastikas painted on her fins. The huge dirigible approached the landing field and gradually lowered herself until she was in the correct position above the mooring mast.

She then switched off her engines and hovered gracefully whilst the handlers below prepared to catch her mooring ropes. She had slowly descended towards the mast until she was just about seventy-five feet from the ground when suddenly someone noticed the fabric near the tail of the ship was flapping loose. As the friends and relatives of the passengers watched, a tiny spark appeared on top of the *Hindenburg*. Then without further warning the whole airship exploded in a solid wall of flames! The burning wreckage fell to the ground almost immediately. The horrified ground crew stood rooted to the spot for several seconds and then ran for their lives as the dying *Hindenburg* came hurtling towards them.

Panic-stricken passengers and crew began leaping from the windows and hatches as the remains of the ship settled slowly onto the ground – the whole thing had taken only forty seconds! In spite of this sixty-two people managed to escape from the burning tomb although some of them were seriously burned. No one ever established for certain what triggered off the hideous accident but the real culprit stood out as being hydrogen gas. The end of the *Hindenburg* also sounded the death-knell for the *Graf Zeppelin* who was quickly returned to her shed and never flown again.

The end of the Zeppelins marked the end of airship travel and also the end for Hugo Eckener. He retired a broken man, only glad that Count Zeppelin himself had not been there to see the tragic end of his creation.

## Chapter Two

## REGGIE WARNEFORD —
## FIRST OF THE GIANTKILLERS

On Sunday, July 28th, 1914, a Serbian student shot dead the heir to the Austrian Empire, the Archduke Ferdinand, and the uneasy peace that had existed in Europe was suddenly shattered. By August, Great Britain found she had been pulled into a world war in order to defend Belgium from Austria's ally Germany.

This was a situation that the British people had been desperately hoping to avoid. For several years they had been reading of Count Zeppelin's spine-chilling invention with horror. In the blurred photographs printed in the newspapers the military Zeppelin looked like something out of a nightmare and it took little imagination to see flocks of these sky-vultures sweeping on Britain and ripping her cities apart with their deadly explosives.

Until now Britain had been protected from enemy invasion by the sea which surrounded her. An island fortress, she had developed her navy until it was the most powerful in the world and before the Zeppelin was invented she was ready to take on anybody. But against aerial attack she would be helpless.

The Zeppelin could switch off her engine and drift in silently on the wind quite unnoticed by the watchers on the coast. She had only to drop her evil cargo of bombs wherever she chose, rise above the cloudbanks and make an unhurried journey home. The British guns had been built to fire at the enemy on land and sea, not monsters in the sky and even if they could be lifted to the right angle to take aim, very few of the guns had enough power to hurl their shells to the height of the elusive dirigible.

The development of heavier-than-air craft (aeroplanes) was speeding up in Britain but the planes were still tiny, fragile things with engines likely to cut out at the worst possible moment and the weapons carried on board were unreliable and primitive. So to the British people Count Zeppelin became an evil bogey-man comparable only with Count Dracula and his invention became even more feared than the newly-invented U-boat. City dwellers lay awake at night waiting for this new menace to come snapping and snarling out of the night sky to destroy them in their beds!

Yet several weeks passed and the raiders did not come. The mood of the people became more optimistic, but then one evening a sea-plane pilot spotted a Zeppelin apparently patrolling peacefully near the East Coast. Straight away lighting restrictions were brought into force. All windows had to be blacked-out before lamps could be lit and street lights were either switched off or heavily shaded. The lights inside buses and tramcars were dimmed and so were the spirits of the people.

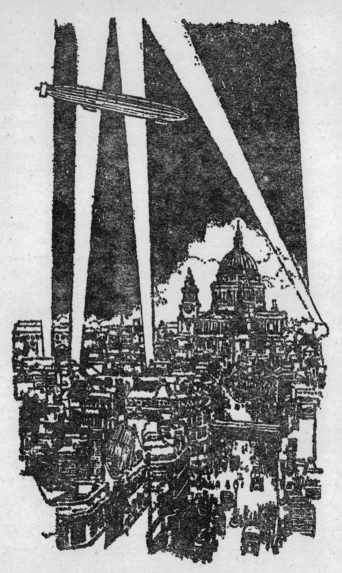

A Zeppelin raid on London

The first raids came in January 1915 when two airships, the L3 and L4 dropped bombs on the undefended resorts of the Norfolk coast, killing five people and injuring thirteen others. This first raid was really a trial run and throughout the spring and summer the airship attacks worked towards a major effort, a full-scale attack on London. The first batch of Zeppelins could not travel as far as the capital so new sheds were erected in Belgium from which London would be just a short hop across the Channel. By May another airship, LZ38 had reached London and dropped 89 incendiary bombs and 30 grenades from Leytonstone to Shoreditch.

Surprisingly only seven people were killed but numerous small fires broke out over a wide area. The vivid glare of the flames gave the impression that whole districts had been burnt to the ground and, as the newspapers of the time were unable to print the truth in case it was read by the Germans, wildly exaggerated rumours began to circulate through the country. This caused even more panic and hatred of the dirigibles.

As the raids continued the British people became obsessed with the idea of destroying the Zeppelins. If only one could be brought down in flames it would do a great deal of good for the morale of the people! Squadrons of Royal Naval Air Service planes were gathered on both sides of the Channel with the sole idea of catching one of the dirigibles unawares. Three Zeppelin raiders were out over England on the night of June 6th. They were LZ37, LZ38 and LZ39. They were spotted by the Admiralty observers and word was sent around to the flying squadrons, who were always on alert. One of these waiting squadrons was situated at Dunkirk in France and amongst its pilots was a twenty-three-year old officer called Reggie Warneford, who had only just finished his course at flying school.

Reginald Alexander John Warneford was born in India of Yorkshire parents and had spent a lot of time travelling the world with them. He was serving as an engineer with the Indian Steam Navigation Company when war broke out. He resigned as soon as he heard the news and came to England to join an infantry regiment. After serving there for a short time without seeing any action Warneford applied for a transfer to the Royal Naval Air Service where he was assigned to Number 2 Naval Squadron at Eastchurch.

Whether Reggie would make a good pilot was a matter of dispute. His best friends agreed that he was far too boastful and swaggering considering his doubtful flying ability. He was inclined to be impulsive and rash, but he had an engagingly cheeky personality. His superior officers soon decided that a bit of enemy action might damp him down a bit so he was let loose on the Huns in an ancient Voisin biplane.

On his very first flight, with an observer on board, he spotted a German observation plane and after ordering the observer to fire on it, he chased it all the way back to its landing field. When, as often happened, the gun jammed, Reggie insisted on climbing out of his own cockpit into that of the observer to try to put it right. The antics of the plane without anyone at the controls can be imagined and it is enough to say that the poor observer had to be helped out of the plane when they landed and had just enough breath left to swear never to fly with such a madman again!

Warneford was then provided with a new monoplane, the Morane Parasol, and allowed to do his worst over the German lines. He had many hair-raising adventures, shooting down enemy planes and strafing gun-emplacements until his plane was so riddled with bullet holes that his commanding officer had to reluctantly provide him with another.

When the three airships were sighted on the evening of June 6th, Warneford was one of the four pilots who were sent up after them. Warneford and a fellow pilot named Sub-Lieutenant Rose took off in single-seater Moranes at about 12.30 AM to chase after the enemy whilst the two other pilots went off to bomb the new airship sheds near the Belgian capital of Brussels. It was the first time that Reggie had flown in the dark and after negotiating banks of thick mist he soon found he had lost the rest of the group.

He searched around for them but there seemed to be no sign of any other aircraft in the sky. By this time it was about 1 AM and Warneford was a few miles west of Ostend on the Belgian coast. The Morane's engine was very noisy and it was difficult to hear any other sound above it so Reggie was hardly surprised when at 3,000 feet he suddenly saw the blue-yellow glare of another aircraft's exhaust flame just ahead of him. At first he thought it was one of his squadron mates, but as he drew nearer and gradually picked out the shape of the craft he realized that she was about half a mile long. His stomach turned over: he had actually found one, it was a Zepppelin, the L37!

In the meantime Wilson and Mills, two of the other pilots from the Dunkirk squadron, had managed to stick together through the fog and made straight for Brussels where they found their target without difficulty. Wilson was caught in the glare of the searchlight and with great presence of mind signalled to the Germans with a torch he was carrying. Thinking he was a German aircraft coming in to land they ignored him until he had released his rack of bombs onto the roof of the giant airship sheds. Mills followed him in and between them they destroyed a shed and an almost new dirigible, the L38, and then got clean away.

The 521-foot L37 was commanded by Oberleutnant von der Haegan and carried 28 crew plus 4 machine guns and normally a large quantity of bombs, which by now had been dropped on England. Reggie's amazement increased as he drew nearer to the underbelly of the mighty bulk, and even by turning his head from side to side he could not see where she began or where she ended! How could such a creature be kept aloft? It was like a floating city!

Before he had time to think further about the matter the Zeppelin's rear gunner opened fire on him and once more his tiny plane was riddled with bullet holes. As Reggie had no gun himself he wisely put himself out of range as speedily as possible. The L37 then began to rise rapidly, having dumped overboard a quantity of water ballast, until she was at about 7,000 feet, well above any height that the little Morane could aspire to. This was when most people would have given up but Reggie was not like most people, and as he saw his big chance floating disdainfully above him he resolved to get the Zeppelin or die in the attempt.

Von der Haegan was glad to be rid of the annoying little hornet and proceeded on his way towards his base at Ghent. However, high altitudes are extremely cold and uncomfortable to fly at so once he was sure he had lost the tiny plane, he began to descend gradually, a few feet at a time, until he was almost as low as six thousand feet. It was now 2.25 AM and seeing the city of Ghent through a break in the clouds, the Zeppelin commander decided to make a bolt for the safety of home.

This was just what Reggie had been waiting for, as he had stalked the L37 for the last hour and a half. He pushed his Morane upwards as hard as he could until he was finally above the descending dirigible and, cutting off his engine, he glided down as close as he dared to the roof of the mighty airship. He pulled out the incendiary bombs that he carried and bounced them one at a time onto the skin of the L37. The sixth one scored a direct hit and there was one immediate, massive explosion. Savage flames a hundred feet high roared out of the top of the wounded airship and Warneford only just escaped being burnt alive. Putting the aeroplane into a dive, he hurtled through layers of choking smoke whilst being bombarded by pieces of burning debris.

The remains of the Zeppelin plunged earthwards and landed on the Convent of St Elizabeth in a suburb of Ghent. Everyone was killed except the coxswain who, as the story goes, fell through the roof of the convent

'The sixth one scored a direct hit . . .'

inside the forward gondola and was thrown out unconscious onto an unoccupied bed. In the meantime the pilot of the British plane was also in difficulties. The shock of the violent explosion had been too much for the Morane and it was all Warneford could do to prevent the engine from stalling. He knew he was well inside enemy territory but he had no alternative but to land her before she took matters into her own hands and dropped to the ground like a stone.

He landed safely and unnoticed in a field that was fortunately shielded by trees. Brooding dismally on the prospects of a long stretch in a German prison camp, Reggie decided to check over his aeroplane once more, before destroying it and trying to escape. To his surprise he found the trouble easily, the length of fuel pipe that ran from the tank to the fuel pump had broken. Searching the plane and his pockets desperately, he produced a cigarette holder, and using the wider end to make the repair, he bound it onto the pipe with the torn-up strips of his handkerchief. After this it took him about twenty frantic minutes to get the engine re-started and it was the most comforting sight in the world to see the propeller once again efficiently spinning.

At 3.15 A M when the first grey fingers of dawn began to poke through the trees, Warneford taxied his loyal plane over the field and off the ground and set off towards where he thought the allied lines ought to be. Once more he found himself lost in the banks of early morning mist but he cruised around hopefully until he came to a break in the cloud and seeing a likely-looking stretch of sand, went in to land. It was Cape Gris Nez, ten miles from Calais and fortunately for Warneford, in friendly France. He was given a cheery welcome and more petrol and by 10.30 AM he had joined his squadron at Dunkirk.

By this time the news that a mighty Zeppelin had been shot down in flames had come through from Ghent and Reggie was happy to take the credit for it. Suddenly he was the hero of the hour. He had given the Allies their first glimpse of success during the Zeppelin peril and proved once and for all that the giant ships were very vulnerable. Within a few hours his name, his picture and his life-story were being flashed across the newspapers of the world and he soon took on the status of a film star. The following day George V awarded the Victoria Cross to Warneford by telegram and the French Government also gave him the Legion of Honour, but poor Reggie had little time to enjoy being a world-wide celebrity.

On June 17th he was sent to Paris where he was to receive his new decoration and collect a new Henri Farman biplane. Paris was just the place for this conquering hero and he loved every minute of it. He was wined and dined and smothered in flowers and was reluctant to leave. Next morning he took off on a test flight in the new Farman, taking with him as a passenger an American journalist, Henry Needham who wanted to do a story on the Zeppelin-killer. Warneford had been instructed to handle the new untried plane with great care yet neither man bothered to strap himself in.

Reggie took off and rose to a height of 700 feet then made a right-handed turn too sharply. The Farman's tail snapped off and the rest of the machine turned over on its back. Both men were thrown out and killed. Reggie's body made a hole in a cornfield two feet across and one foot deep. The Cross of the Legion of Honour, which he had just received had been driven right through his tunic and into his chest. A French journalist wrote at the time, 'He who defied the storm has been killed by a breeze.'

## Chapter Five

## MATHY AND THE
## ZEPPELIN RAIDERS

| Zeppelin flieg | Zeppelin, fly |
| Hilf uns im Krieg | Help us in the war, |
| Fliege nach England | Fly against England |
| England wird abgebrannt | England will be destroyed by fire |
| Zeppelin, flieg! | Zeppelin, fly! |

As the attacks by Zeppelin raiders on Britain began to get under way the German Navy rapidly expanded its Airship Division. Several thousand men were assigned to this section, three new bases were built and the Zeppelin factory began to churn out bigger and better military airships as fast as they could. These airships were more streamlined, with a control cabin gondola that fitted snugly against the airship's belly to cut down wind resistance. They had larger, more powerful, engines and the novel addition of a 'sub-cloud' car.

This invention was a tiny fish-shaped car with a tail-fin that was let down from the main body of the airship on two or three thousand feet of steel cable. It was designed to enable the large bulk of the Zeppelin to remain hidden in a bank of cloud whilst a brave volunteer would be lowered as much as half a mile from the airship. When he had a good enough view of the lights beneath him he could report the information back to the airship by telephone so that the navigator could work out the airship's exact position.

Commander Peter Strasser, a great admirer of the old Count and a dedicated Zeppelin enthusiast, was now Chief of the Naval Airship Division and he trained his rapidly multiplying airship crews with skill and vigour. By the summer of 1915 the Airship Division had built up such a name for itself that Strasser succeeded in getting the Kaiser's permission to raid London itself with his Zeppelin squadrons.

Earlier the Kaiser had been very much against the bombing of London. King George V was his cousin and the idea of destroying his own family's palaces troubled him. By the summer of 1915, however, the Airship Division had gained such prestige with successful raids on other parts of England that the Kaiser was forced to give in. At last the eager Strasser had permission to launch his airships onto the British capital.

As a result of the earlier air raids outside London the airship captains were already the toast of Germany and of these men one stood out as being the bravest and most resourceful of all. Commander Heinrich Mathy was young, bold, classically handsome and completely ruthless. Cool-headed in a crisis, he was highly skilled in all aspects of airship flying and navigation and his fanatical duty to the German Fatherland enabled him to kill totally without emotion. He was born in Mannheim and entered the German Navy in 1900 where he served for several years in destroyers. In 1913 he transferred to the Naval Airship Division and soon became a reliable dirigible captain. He first commanded LZ9 and then took over LZ13 for its first raid on London.

Four airships set out across the Channel on September 7th but they were soon reduced to three when one developed engine trouble and had to turn back. Two of the three, the L13 and L14 were heading for the Haisboro' lightship anchored twelve miles off the Norfolk Coast. For five hours the airships droned on over the grey misty sea without sighting even a fishing boat. At last they reached Haisboro' where they were fired on by British guns. Rising disdainfully out of range, the two Zeppelins continued on their course towards London. Mathy sent the men to their action stations, the bombing officer and his party to the swaying catwalk in the middle of the ship and the machine-gunners to the control car.

The 'top gunner' had the most unpleasant job of all; he had to climb a ladder through the dark body of the airship, buffetted all the while by the large, smelly bags of gas. His machine-gun post was a little niche in the outer envelope of the airship, totally exposed to the wind and rain. He was supposed to sit there and look out for aeroplanes but often he was too cold and frightened to think of anything but getting down again.

Although the countryside was blacked out, Mathy had no difficulty in following the rivers and the railways to London itself. However careful the Londoners were about the blackout, the Thames always shimmered and gleamed in the darkness and from its various twists and

A top gunner

bends a good airship commander could pick out any point in London. As the L13 flew slowly over north-west London a thin finger of blinding light reached up from the ground and began to probe the sky for him. More searchlights began to join it and suddenly it seemed as if London had turned into a giant octopus reaching out its tentacles to strangle the invading air-ships. As the searchlights spotted their target, there came the red flashes and roars of the anti-aircraft guns but as yet they were not accurate enough.

Mathy began by bombing around Euston Station,

going from there to Tower Bridge dropping incendiaries all the way. Fortunately he missed the bridge itself but dropped many bombs over Holborn, Smithfield Market, the Guildhall and on the doorstep of the Bank of England. He then went on to drop bombs on Liverpool Street Station, two of which fell on omnibuses killing and injuring many people. He dropped the rest of his deadly cargo over the City of London itself and then turned for home.

Londoners were appalled when they crept out to inspect the smouldering trail of destruction that Mathy had left behind. People rushed here and there searching the wreckage for missing relatives, while many thousands of pounds worth of property went up in smoke. It was the kind of catastrophe that Great Britain had never seen before and a grim pall of silence settled over the capital. The raids continued but the defenders, knowing what to expect, grew more crafty. There were 26 anti-aircraft guns in action now and more and better guns were being added all the time. On September 13th an airship was hit by anti-aircraft fire and had to return home. On October 13th, during an airship raid led by Mathy 71 people were killed and 128 injured. But in spite of this it was for the British the beginning of their long fight back, because the airship crews returned to Germany after that raid with their confidence rather shaken.

The searchlights over London had now been increased, the ground guns seemed to be firing much more accurately and, although the Germans did not realize it, the British were now using a new explosive shell that could be fatal if it made a hit. Four aeroplanes had also been added to this formidable defence and although they did not damage any airships that night they were obviously getting in plenty of practice. For the first time the Zeppelin raiders felt that the British would not take this beating lying down.

The raids continued but even Commander Strasser had to admit that new difficulties were constantly appearing. Fighter planes emerged suddenly from behind clouds, larger and more powerful guns barked at them not only from the land but also from the sea. An incendiary machine-gun bullet had now been invented that set fire to the huge gasbags immediately on impact. The mighty dirigibles were at last being brought down!

On September 23rd, ten days after Mathy's spectacular raid on London three of the new super-Zeppelins L31, L32 and L33 set off for London once again. Mathy commanded the L31, with Strasser along for company, and they successfully bombed Mitcham, Streatham, Kennington and Walthamstow and made a safe escape over Great Yarmouth.

The L33's captain, Böcker, also started his evening well, steadily bombing his way from Billericay in Essex to Wanstead. He then began on East London itself. By now he was under a heavy barrage of fire from the ground guns and from Wanstead onwards Böcker was aware that besides having a smashed propeller L33 was losing gas at a fairly rapid rate. He turned and slowly headed for the sea but now, to add to his troubles, he was being hotly pursued by an aeroplane. The pilot managed to let off two drums of ammunition into the stern of the airship before he lost sight of her. He need not have bothered for the L33 was already doomed. She was sinking fast and although the crew were throwing over the side anything that wasn't actually bolted to the floor it did not seem to make any difference. Obviously they could not get her across the Channel in this state, so Böcker landed her in a field about three miles inland, and then set fire to her.

Determined to do everything properly, as they were now at the mercy of the British, Böcker knocked on the nearby cottage doors to warn the locals of the inferno on their doorsteps. Naturally enough, no one answered, they were much too scared! When the airship had burned out Böcker fell the crew in and the tiny army marched off to find someone to whom they could surrender. This unfortunate person turned out to be a local constable pedalling home on his bicycle. With great courage he ordered the Germans to 'Come along with me' and to his surprise they very politely did so. They were eventually put under guard in the village hall and given breakfast before being taken to Colchester by motor lorry.

The third Zeppelin, the L32 commanded by Peterson, crossed the Thames at Dartford and flew into a web of searchlights which successfully held her until a fighter pilot, Fred Sowrey, set light to her with a continuous stream of machine-gun fire. The burning Zeppelin fell to the ground where her molten skeleton was found impaled on an old oak tree like the bones of a prehistoric monster. There were no survivors.

The wreck of the L33 was put under a heavy guard

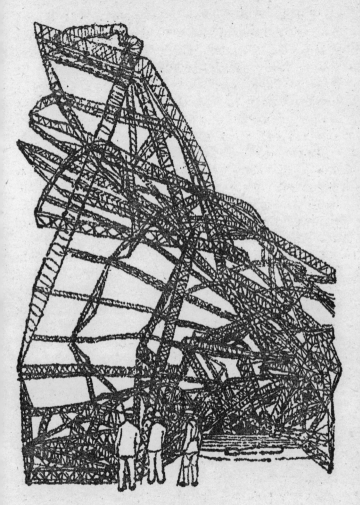

The wreck of the L33

because, as her structure was almost completely intact, she was particularly valuable to the experts. Everything was removed that was considered important and the field was then opened to the public at the modest admission fee of one penny per head. In this way £80 was collected for Red Cross funds. It had been a great night for the British; the new aeroplane had shown itself to be a valuable defensive weapon and was beginning to knock the stuffing out of the Zeppelin squadrons. Strasser was worried at this turn of events and ordered more caution on the part of his men. London must be left alone as the strength of its defences made it too dangerous. The Zeppelin crews must find other important targets in England.

The decision came as a relief to the airship crews, particularly Mathy's men who had witnessed Peterson and his crew being burnt alive. They had flown 120 successful missions and felt that their turn must come soon, particularly as their captain always managed to find his way to London and to expose his airship to the maximum amount of danger.

They were given a few days' well-deserved break from flying due to unsettled weather but by October 1st, Commander Strasser decided the time had come to hit England again. The crew of the L31 resignedly prepared their ship and their hearts sank still further when their captain, Mathy, announced his intention of having one last crack at London. He was their captain, and they had no choice but to go with him into the mouth of the British lion.

On moonless nights the Zeppelins were always kept

ready for action, the great hangar doors open wide and 'walking out' parties waiting on call. On October 1st the gas bags were filled at dawn, leaving an unmistakable sickly-sweet smell everywhere, and the ships were fully stocked with petrol, machine guns and ammunition. The bombs were then loaded on by the maintenance crew. They had to be hoisted into the middle of the airship by means of tackles and suspended by hooks to the bomb frames ready to be despatched onto England.

When the 'bombing-up' was completed the crews put on their warmest clothing and their life-jackets and collected flasks of piping-hot coffee before climbing on board for the 'weighing-in'. Before a ship could leave, a check had to be made on her 'positive lift' and enough water ballast had to be discharged to make sure that the airships would not rise higher than necessary or creep along too near the ground.

As soon as everything was completed on L31, Mathy gave the go-ahead. The handling party then took hold of the handrails running along the bottom of the gondola and gently pulled the swollen bulk out into the night air. As Mathy shouted 'Cast off' the men below let go the ropes that held her to the earth and after a moment's pause the L31 surged upwards and was gone.

Of the eleven ships that started out on that mission on October 1st, three never even reached England. Five got lost over Lincolnshire and Norfolk and were unable to find any suitable targets on which to drop their bombs. The L21 became so thickly encrusted with ice that she had to jettison her bombs as if they were ballast before returning home. The captain could not even make radio contact as the ice-coating on his aerial was three inches thick.

Although the airship raiders had a glamorous image, on nights like this one could appreciate what a dirty, unpleasant job it was – particularly for the mechanics. For five hours they would have to huddle next to their engines fighting fear, nausea and fatigue. The only bit of variety was when, at very rare intervals, they were allowed to go up for a hot drink. In order to do this they had to climb into the main body of the ship and walk along half the length of the freezing catwalk with only the reeking gas bags for company. They would then have time for a few quick gulps of coffee before returning to their posts beside the engines.

The control car crew did not have things all that easy either. Although they were not subjected to the boredom and fumes of the engine room they had to endure the discomforts of sub-zero temperatures in the icy control cabin. In fact conditions were very uncomfortable for all concerned, even when things were going smoothly, but when an airship came under attack it was like a nightmare. The men had to perform their normal duties sliding up and down the catwalk inside that vast swaying cave whilst the blinding flashes of the shells reminded them that any moment they could be caught in the middle of a plummeting inferno. The mental and physical exhaustion of these men must have been great!

Mathy's L31 reached England at 9 PM coming in over Lowestoft. Whilst the others were floundering around looking for targets Mathy's unerring eye picked out the Great Eastern Railroad which led him straight into London. Usually he would have been able to fly straight into the centre, drop his bombs and be away, but tonight outer London's searchlights picked him up immediately, and, twisting and turning to avoid them, he had quickly to veer off to the North. The defenders of the city were becoming too clever, even for him!

Yet where other airship captains knew when to give up Mathy did not, and bobbing and weaving, diving and climbing, he began circling the north-eastern fringe of the city, determined at all costs to break through the line of defence. He just could not do it, for however carefully he moved the British heard him, aided by special sound detectors shaped like enormous megaphones, and once again the searchlights bathed him in a blinding light.

This great hue and cry could not fail to be noticed by the four aeroplane pilots who were in the air at the time, waiting for just such an opportunity to get to grips with a Zeppelin, particularly one piloted by the mighty Mathy. One of them was flown by Second-Lieutenant Joseph Tempest, who had been at a dinner party when an urgent phone call had sent him aloft to look for the airship that was reported to be approaching London. He was about fifteen miles away when he noticed that the searchlights somewhere north of London were forming a huge pyramid of light near to which he could see a tiny cigar-shaped object.

He headed across London and straight for the Zeppelin but as Mathy saw the plane swooping towards him he immediately began to climb out of range. However, fighter planes had improved in the last few months and Tempest could also play that game. Before Mathy could get any higher the little plane was above him and diving down onto him, firing as she came. His first burst of fire went into the Zeppelin's side and then, banking his machine, Tempest fired another round under the hull. As a finale, he flew down the whole length of the ship from the tail to the nose pumping her full of bullets.

He was still firing when he saw a rosy glow begin to radiate from inside her. The pink glow turned to a fiery red and she lit up the night sky 'like an enormous Chinese lantern'! The flames then shot out of the nose and began to lick greedily round her and Tempest knew that he had made a hit. The flaming monster reared upwards about 200 feet and then, roaring hideously, fell swiftly to the ground almost taking the fighter plane with her. Tempest had to put his aircraft into a spin and he just managed to corkscrew out of the way before the airship brushed past him. Feeling sick and exhausted, the fighter pilot made his way back to base with great difficulty. By now it was very foggy and Tempest was shivering so much, probably with shock, that somehow he managed to crash his plane on landing, but apart from cutting his head slightly on his machine gun, he escaped unhurt.

The L31 ended up in two separate piles in a field near Potter's Bar. The forward half of the framework was tangled around an oak tree, whilst the rear half lay scattered over a wide area of land. In spite of the drizzle the fuel oil that had been scattered everywhere caused the wreckage to carry on burning for several hours.

Some of the braver villagers came out to see if they could help and some distance away from the wreckage found a German officer who was still breathing although his body was so badly broken that he had obviously leapt out from a great height. He was still dressed very correctly in his heavy uniform overcoat with a woollen muffler around his neck. As soon as they tried to move him he died. It was Mathy. The bravest airman of all, the hero of all Germany. He had made his last trip to London. His spectacular death marked the end of the Naval Airship Service and Commander Peter Strasser never recovered from the shock. Never again was London raided by the Zeppelins.

Chapter Six

# HORROR ON THE ICE PACK

The tragic story of the airship *Italia* began in 1922 when a young Italian aeronautical engineer, Umberto Nobile, designed a successful series of dirigibles. Unfortunately 1922 was also the year when a new political party, the Fascists, swept into power led by the powerful and notorious dictator Benito Mussolini. The Fascist Government, who were on the lookout for money-making schemes, immediately made plans to take-over Nobile's aircraft factory against his will. Nobile, a determined man, held out as long as possible but eventually was forced to sell his plans to the Government and was taken into the army as a lieutenant-colonel.

Nobile was allowed to continue with his designs and in 1924 the N1, a new semi-rigid airship, was completed. She was beautifully streamlined, yet shorter than most of the dirigibles being built at the time, a mere 348 feet. Her greatest asset was her keel which, built like the flexible human backbone, could bend and twist in high winds without breaking. Nobile was delighted with her and at once began to make plans to prove her worth.

At the same time that Nobile was putting the finishing

touches to the N1 airship, Roald Amundsen, a Norwegian explorer, was attempting to fly over the North Pole in a sea-plane. When this attempt was unsuccessful, he looked around for an alternative method of flying. His eye fell on Nobile's new creation and he decided to approach the Italian Government with his ambitious plan. The Fascists, fearful that the adventure might end disastrously and show Italy in a poor light, were reluctant to get involved so instead of cooperating they sold the airship to Norway for 75,000 dollars. The N1 was re-christened *Norge* and Nobile was hired as chief pilot for the expedition which was to be captained by Amundsen and financed by an American, Lincoln Ellsworth.

In May 1926 the *Norge* left its hangar in King's Bay for the long and hazardous journey to the North Pole. On her first day out she was enveloped in dense fog, and soon thick ice coated the outside of the ship, yet she plunged on through the elements hardly seeming affected by them. Eventually she reached the Pole and staggered on towards Point Barrow in Alaska, growing heavier and heavier as her ice coating grew thicker. Just as she reached the coastline of Alaska, she was caught in a howling gale and forced to land in an Eskimo village not far off from her original destination. The polar expedition was hailed as an outstanding success; the airship and her intrepid crew were acclaimed throughout the world and Nobile and Amundsen returned to their respective countries covered in glory.

However, no one realized at the time the difficulties the two men had encountered, not just working against the elements, but also working with each other. From the beginning Nobile had felt bitter about being made second in command on his own airship and as Amundsen persisted in treating him as an inferior hireling, the resentment grew. By the end of the voyage a full-scale feud had developed and even after the two men returned home they continued to attack each other in newspapers and magazines.

Nobile, now promoted to General, was determined to prove that he could get to the Pole again without any help from Amundsen. He immediately set to work to design an updated version of the *Norge*. Mussolini, anxious to grab any glory that was going, agreed to finance another polar expedition but this time with an all-Italian crew.

The new dirigible, called the *Italia*, had several important improvements. The control car was more spacious and comfortable, and a different type of covering had been used for the envelope to make it both stronger and lighter. In addition to the Italian crew, Nobile planned to take with him two scientists from the original polar flight, Finn Malmgren, a Scandinavian, and a Czech professor, Dr Francis Behounek. The General, determined to leave nothing to chance, packed the dirigible with rifles, pistols, sledges, rubber boats and snowshoes in case they decided to land. Their provisions were also carefully planned and varied, the main food being a dried meat and vegetable mixture called pemmican.

At 1.15 AM on April 15th, 1928, the ghostly form of the *Italia* floated eerily skywards from Milan on the first

leg of her arctic trip. The date had been fixed to co-
incide with good weather forecast by the meteorologists,
but for some reason the good weather never appeared.
Instead the *Italia* ran into violent weather right from the
start. It began with rainstorms and was rapidly followed
by hail, snow, fog, heavy winds and lightning – just the
sort of weather that had wrecked previous airships. The
snow-battered *Italia* gamely ploughed on for ten days

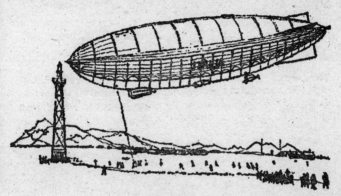

The *Italia*

until her propellers became so badly splintered and
chewed by hailstones that she had to come down in
Germany to make repairs and to take on extra fuel and
ballast. Several days were needed for repairs before she
was fit to carry on. She reached King's Bay on May 6th,
only a little later than the *Citta Di Milano*, an old Italian
ship sent out to be their base camp for the duration of
the expedition.

The *Italia* began her series of polar investigations

some days later. Although dogged by continuous bad weather, Nobile and his countrymen were delighted at the new discoveries they were making. On May 23rd Nobile decided that they were ready to try a major assault on the Pole itself. In preparation each crew member left the base camp that day wearing thick, woollen underwear, a woollen suit and a lambswool hooded anorak and trousers. They also carried watches, compasses and dark glasses to protect their eyes against the glare of sun on snow which can cause snow-blindness.

Glistening and shimmering in the bright Arctic sunlight on that fateful day, the *Italia* must have looked a strange, futuristic machine indeed compared with the laborious dog sledges that were used by the explorers of the day. The journey to the Pole was a happy one, clear blue skies staying with them almost all the way. They finally reached their goal early on May 24th and this time the Italian flag alone was dropped onto the ice, followed by a large oaken cross given to Nobile by the Pope.

For the Italian crew and Nobile in particular it was a proud and emotional moment and the radio operator immediately sent off a message to Rome telling of their success. The engines were then switched off and whilst the howling Arctic wind pulled and tugged at the fragile frame of the airship they all listened whilst a popular Italian song, *The Bells of St Guisto*, was played on a portable gramophone. This was followed by a celebration drink for the crew which served to warm everyone up and generate an even happier atmosphere on board.

It was a very lighthearted crew that swung the airship away from the Pole two hours later and into disaster.

As they turned south for home there was an immediate change in the weather. The friendly wind that had helped to blow them north had greatly increased its strength and, as they turned back, it fought them every inch of the way. Suddenly they heard a series of cracks like pistol shots which turned out to be chunks of ice on the propeller blades being flung like bullets through the delicate fabric of the envelope. Nobile sent men to repair the holes but as they did so fresh holes appeared. This hair-raising state of affairs continued from 7 PM that night until 3 AM the following morning by which time both the *Italia* and her crew were showing signs of great exhaustion.

By then savage winds were buffeting the huge bag from all directions and, unable to make any forward progress, she began bucking and rolling like a frightened stallion. Even when there was a slight lull in the storm the crushing weight of her heavy coat of ice only allowed the stricken giant to stagger on at a snail's pace. Nobile, his face white and drawn through worry and lack of sleep, realized that they were now in real trouble. With King's Bay 350 miles off and the storm showing no signs of abating only a miracle could save them! There were to be no miracles that day. Suddenly the stern of the ship began to list badly and the cry went up, 'We are heavy!'

Nobile stared with horror at the instruments in front of him and realized that they were falling at the rate of two feet a minute.

'Full ahead all engines! Emergency!' he shouted. As this had no effect at all he then ordered 'Up elevator' which served only to lift the bow of the ship with the stern still sinking alarmingly rapidly. Finn Malmgrem, peering through the window in the nose of the control cabin broke the shocked silence. 'Look,' he shouted, 'There's the ice-pack!'

Pressing their noses against the ice-encrusted windows the crew saw with horror the pointed fingers of the ice-pack hurtling up to meet them as if on the surge of a tidal wave.

'Stop all engines. Close all ignitions!' A crash was now inevitable but the risk of fire from the giant bag of hydrogen was very great so all possible precautions had to be taken. Nobile thrust his head through a hole in the pilot cabin and watched as the tail engine of the wounded ship smashed into the unwelcoming ice pack.

'God save us,' Nobile muttered as he braced himself for the colossal impact. As if in a nightmare he felt himself being flung backwards and forwards against the framework of the cabin and heard the snap of his own leg breaking like a twig. Then he slipped quietly into unconsciousness.

Nobile was the most seriously hurt with two broken limbs and fractured ribs. As he lay prone on the ice the crew members who could still stagger, picked their way dazedly out of the mangled wreckage. The control cabin gondola had been completely severed from the rest of the airship. The dark shadow of the *Italia*'s envelope still towered over them and as the survivors looked up at it they caught a glimpse of white, horror-struck faces staring down at them. Six crew members who had been travelling in the bag of the airship were now marooned in it. As their eyes met, the bag of the airship began to drift upwards again.

One of the men, quick to realize the urgency of the situation, began to throw overboard fuel and provisions for the men down on the ground. It was a brave and inspired act for immediately afterwards the derelict balloon was caught in a fierce gust of wind and, completely out of control, was whisked away into the mists. The *Italia* and her six helpless crew were never seen again!

The first day on the ice-pack was a nightmare for everyone, but particularly for Nobile. He lay on the ice for hours delirious with pain whilst Mariano his second in command did his best to gather the pathetic little band into some sort of order. By the end of that day an emergency camp had been struck. A tiny tent that had been thrown from the *Italia* was stretched to accommodate all nine of the survivors, and the General's dog Titiana. Apart from Nobile only one other person had been badly injured, Cecioni, who also had a broken leg. Their only moment of joy was when they found that their radio was completely unharmed by the crash. Every hour they sent out a message to the *Citta Di Milano,* 'S O S S O S.'

There was no reply from the ship.

The scattered parcels of supplies were collected from where they had fallen. There was pemmican, chocolate, malted milk, butter and sugar. With care it looked as if they might be able to last out for as long as forty-five days. They also discovered some 'altitude bombs'. These were glass globes full of red dye used by Nobile to tell how far the *Italia* had been above ground. The dye was then used to paint the 8 foot by 4 foot tent a bright red, in the hopes that it might be spotted more easily. The bedraggled aeronauts then huddled under the only blanket they could find and waited for help. It was to be a long time coming.

In spite of the excellent condition of the radio and the fact that they could hear the *Citta Di Milano*'s messages loudly and clearly there was absolutely no response to the *Italia*'s SOS. The base ship, for some incredible reason was just not bothering to tune in to their wavelength! Finally on the fifth day of their ordeal they heard in a radio speech given by the Italian Senate that the *Italia* had been lost and her crew given up for dead! The nine survivors in the red tent were filled with bitter anger and frustration.

The *Citta Di Milano* was an aged hulk of a ship, inadequately equipped for the job she had been given as the *Italia*'s base ship. She was not even fitted out as an ice-breaker and her engines were old and frequently broke down. Her captain, Guiseppe Romagna seemed to be well chosen to command such a useless ship. From the beginning of the enterprise he had taken little interest in the *Italia* and had not made the least attempt to cooperate with her. A timid man, lacking in initiative, he had ignored the last messages from the *Italia* reporting that she was having trouble. He made no attempt to keep a radio track on her although it must have been obvious that things were going badly. His excuse for not listening-in after the *Italia* was known to be down was, 'I did not think the survivors would have any radio, so we did not bother to post listeners!' After several days of complete inactivity Romagna finally radioed Italy to ask if an ice-breaker could be sent from Russia to look for any survivors.

Twenty-four hours after the crash the fierce winds died down and, with the aid of instruments taken from the wrecked control cabin, Malmgrem and Nobile were eventually able to work out the approximate drift of the ice pack. Apparently they had drifted twenty miles in two days towards an island called Franz Joseph Land, a desolate waste where there would be no hope of rescue. They began to discuss the possibility of a march across the ice to another less remote piece of land but with the two badly-injured men it was not considered to be a practical idea. However, as they drifted farther away from help and there was still no message from the *Citta Di Milano* it was decided that the three fittest men, Mariano, Zappi and Malmgrem, would attempt a hike across the ice pack to send back aid to their companions. Malmgrem reckoned that if they marched steadily at about ten kilometres a day they could reach Cape North, an uninhabited piece of land at the far tip of the Arctic islands of Spitzbergen, in about sixteen days. They were to leave the next morning.

That night their sleep was disturbed by the growling of a very large polar bear. Malmgrem took a pistol and crept stealthily out of the tent. Nobile pulled his dog, Titiana, close to him and held her so that she could not bark. The bear towered over Malmgrem and it took a lot of courage and three close-range pistol shots to stop it in its tracks! The great animal was skinned and cut into pieces whilst it was still warm and the following morning they had bear-meat for breakfast. Bear meat turned out to be greasy, tasteless and much tougher than they had hoped but as they now had four hundred pounds of bear it would have to play a large part of their future diet. That evening the little party set off, taking some of the valuable supplies of food and clothing with them. They also took an axe and a knife, leaving the valuable pistol for the immobilized base camp. As they left the wireless operator again sent out his signal: 'SOS *Italia*,' 'SOS *Italia*.' It was followed by a deep silence.

Spirits rose after the departure of the three men; something positive had now been done about a rescue and conditions in the little tent seemed less cramped now that there were three less people. The whole of that week passed with Biagi doggedly attempting to make some contact with the outside world. Finally on June 7th they heard on a news bulletin that a Russian farmer had picked up a message on his private radio from the stranded men! It hardly seemed possible – someone as far away as Russia knew they were still alive but their own nearby *Citta Di Milano* still turned a deaf ear. Surely now something would be done about a rescue. Yet it was two whole days before their base ship called them on her radio!

Only then was Nobile pleased and he gave Biagi a whole bar of chocolate as a reward for his unceasing efforts. For the first time they felt a surge of real hope but it was soon to change to despair. Mariano, Zappi and Malmgrem were still out on the ice and the hopeless *Citta Di Milano* still sat in King's Bay harbour, her boilers empty of steam.

Back in civilization the amazing news of the Italian survivors brought a great burst of activity. Rescue operations were set in motion that were to make polar history. Many countries sent ships, planes and men to look for the survivors. The *Hobby*, a small whaling ship, began to search north of Spitzbergen along with the Norwegian ship *Braganza* and the Russian ice-breaker *Malygin*. Teams of overland rescuers set off in dog-sleds and, most surprising of all, Roald Amundsen, the Norwegian explorer and Nobile's bitter enemy also joined the ranks of the daring aviators who set off to search the polar ice cap. Every day saw a new rescue mission setting off, while another returned (or in some cases, failed to!).

The papers were full of accounts of people who had met with disaster. One of the most lamented of these was Amundsen himself, who disappeared without trace into the Arctic wasteland. His plane was never found. Ironically the only country that made no move to send help was Italy itself.

The first planes were sighted by the inmates of the Red Tent on June 17th. The survivors frantically fired flares but somehow the planes just did not see them and turned away and vanished over the horizon. On June 20th they heard the drone of another aircraft engine. Nobile and Cecioni dragged themselves outside with the others to watch her. She grew larger and larger until they could actually see the head of the pilot in the cockpit.

Rooted to the spot, they watched until he lifted his hand and waved to them and dropped provisions and equipment overboard.

Three days later, on June 23rd, a Swedish pilot, Einar Lungborg, flying a small Fokker ski-plane, glided to a safe and bumpy landing near to the Red Tent. Lundborg and his co-pilot made their way across the ice to the camp and were horrified at the conditions they found there. The gaunt skeletons anxiously staggering towards them bore no resemblance to the gallant airmen who had crashed in the *Italia* four long weeks before. General Nobile, his eyes full of tears, looked particularly ill, and as there was only one spare seat in the tiny plane, Lundborg insisted that the General should take it.

'That is impossible,' said Nobile, determined to stay with his men. 'Cecioni is the weakest, you must take him first!' However, Lundborg refused. He argued that the General would be by far the most useful in helping them to mount further rescue operations. There were still Malmgrem, Zappi and Mariano to be found and also the poor men who had been swept off in the bag of the *Italia*.

Nobile, much weakened by the pain and delirium of the last few weeks, hesitated, and after some persuasion from his men, finally gave in. Along with his little dog, Titiana, he climbed into the cockpit of the Fokker. Soon the Red Tent with its pathetic inhabitants was left far behind and he was reluctantly taken on board the *Citta Di Milano*.

After Nobile's departure, Viglieri took charge of the bedraggled band for the last few hours before Lundborg and the other pilots returned to collect them. His main concern was to get poor Cecioni over the jagged ice to the landing strip. It was a difficult trip and Cecioni got very wet in the process but in a fairly short time he was able to sit and wait with the others. It was another six hours before they heard the noise of the Fokker's engine again. It was obvious, straight away, that the plane was in serious difficulties and Lundborg seemed to be unable to control her. It came as no surprise to the horrified observers when her skis became caught in the ice and pitched the plane over on her back, completely smashing the propellers. Viglieri rushed over to the plane filled with concern for the pilot. His anxiety turned to disgust, however, when he found that the pilot, besides being unhurt, was completely drunk! He had obviously celebrated the rescue before actually finishing the job!

After this last bitter blow, the five survivors pitched their tent next to the aeroplane and, along with the sheepish and now sober Lundborg, they huddled together, discouraged and depressed whilst the hours of waiting lengthened into days. Two weeks later, on July 6th, another plane arrived. The Italians were all in a low, desperate state by this time but Lundborg, the pilot, had gone utterly to pieces. Although physically fit, his severe mental distress was having such a depressing effect on everyone else that Viglieri had no choice but to send him back in the plane's only passenger seat. It was to be the last plane they ever saw.

Over the last few days the weather had become warmer and with the ice pack beginning to melt and break up, it was no longer safe to risk a landing on it. The Swedish pilots, angered because no Italian planes had been sent to help, decided that they were no longer prepared to risk their lives attempting a rescue. After forty-four days on an ice floe, salvation for the men of the *Italia* seemed as far away as ever.

They had reckoned without the battered old Russian ice-breaker *Krassin*! Although dirt-encrusted and obsolete, she still had a good set of engines and, as an added bonus, a crate containing a *Junkers* aeroplane to be used in the search. She lumbered out from Bergen harbour and by June 27th was inside the Arctic Circle. Two days later her bows of reinforced steel began tearing through the thick blocks of ice. For many of the crew new to the game it was a terrifying experience. The ear-shattering explosions of cracking ice and the giant slivers of it that were being hurled onto the ship could not have made it the pleasantest form of travel! In this manner the old vessel buffeted and harrassed her way through the stubborn ice pack until July 7th when the *Junkers* was assembled and prepared for take-off.

The pilot, Chuckhnovsky, after several fruitless missions in the right area, flew off once more on July 10th and failed to return. Just when everyone began to fear the worst he sent a radio message. 'Cannot find *Krassin* in fog,' he said. 'Have discovered Malmgrem trio. Will attempt to land in Seven Islands area.' This caused great rejoicing on board, particularly as it meant the three men should be within twenty miles of the *Krassin*. The great ship was soon on her way again to the soul-destroying tune of hollow steel hitting solid ice. By 5 PM the following day she had reached the location of the three survivors. 'One hundred roubles,' roared the captain, 'for the first man to sight the survivors!'

Almost immediately a pitiful figure could be clearly seen some distance away bending over an even frailer companion. It was Captain Zappi and Mariano. Both men were skeletal, frost-bitten and very dirty. Mariano was particularly ill and near starvation and Finn Malmgrem had been dead for some time. After making the two men comfortable on board, the *Krassin* pressed on. Following Viglieri's instructions over the radio she reached the site of the Red Tent at 9 PM that evening. To everyone's surprise they actually found it sitting there, next to the ridiculous upside-down plane.

After forty-nine days in the wilderness the solid, reliable form of the *Krassin* seemed too good to be true. The haggard airshipmen vacated their tiny prison with relief and were pulled thankfully on board the ice-breaker. Then the mists came down and the *Krassin*, short of fuel, had to turn for home. No other survivors were ever found.

Chapter Seven

## JOURNEY INTO DISASTER

Whilst the attention of the world was focused on the fate of the *Italia* and the ordeals of her unfortunate crew, the British people were preparing to launch their own new airship fleet.

Since the break-up of the R38 over the Humber in August 1921 there had been mixed feelings in Britain about the future of rigid airships. Some people felt that airship building was too risky and expensive and thought it would be better to sit back and watch other countries iron out the problems. Others, however, thought differently. They maintained that Britain had led the world in lighter-than-air travel with her R34 in 1919 when she had been the first to cross the Atlantic. They felt that to stop experimenting now would result in a great loss of world prestige. Also the German Zeppelins had proved a frightening weapon in World War I and in the event of another war the side with the most successful airship fleet might have a distinct advantage. So in 1924 the Government appointed a Cabinet Committee to look into the matter. The result was a decision to launch an all-out airship programme that would decide once and for all whether the airship could be made a safe, commercial proposition.

This new experimental programme, intended to stretch over three years, was to cost the tax-payers £1,350,000 and at the end of it they were to get two new

The R101

super-ships, the R100 and the R101. These airships were to be large enough to carry vast cargoes and were to be fitted out for long intercontinental flights. The Government also decided that to promote competition and keep everyone on their toes, it would be a good idea to let a private firm, the Airship Guarantee Company Ltd, build one airship and the Government aircraft factory build the other. In this way they would get the benefit of two separate sets of brains working on the same problem.

It was planned that both airships would have a maximum speed of 70 mph and be able to lift as much as 150 tons, almost a third as much again as the world's only successful airship, the Zeppelin. Conditions were also laid down that both airships must be able to run on fuel that could be used safely in hot countries and each had to prove to a special research committee that it was strong enough to stand up to the fiercest weather conditions. The R100 was to be built in Howden in Yorkshire by the Airship Company Ltd, to the design of the brilliant engineer Barnes Wallis (who later designed the famous bouncing bomb which blew up the Moene Dam in World War II).

The R101 was to be Government financed and built in Cardington. The designer was Colonel Vincent Richmond who had spent a long time studying airship techniques in Germany. When Dr Eckener, the Zeppelin designer, saw the plans for the R101, he was quoted as saying, 'Very nice – but isn't it a little big?' The Government's policy had all the ingredients of a sporting contest between the rival manufacturers and their progress was followed with interest. The new plan also brought rejoicing to Cardington in Bedfordshire, an area desperately in need of more employment.

From the beginning the R101 seemed to develop more problems and run into more expense than her sister-ship. When the R100 group found their ideas did not work out properly, they could afford to scrap them and start again. The state-financed group, on the other hand, had to account for every penny they used to Parliament and could not afford to abandon anything in case they were accused of wasting the public money. They also had the extra responsibility of the workers and their families at Cardington who depended on them for a living at a time when there were many unemployed in the country. As a result of all these difficulties, R101 had several problems that were never properly solved.

An interesting feature of the R101 was the use of goldbeater's skins to make the gasbags. These were membranes that had originally been part of a bullock's intestines. They had been cleansed of fat and scraped with blunt knives by teams of female workers; the skins were then soaked in glycerine and eventually stretched and varnished before being made into bags. In these days of nylon and polythene it seems laughable to think that without the help of more than a million oxen the R101 could not have left the ground.

The actual assembling of the parts of the R101 took place in her mooring shed at Cardington and it must have been a tremendous sight to watch. Inside the hangar, hundreds of girders were woven like a giant Meccano set into an enormous metal cobweb. There were great ferris wheels 400 feet round to be slipped into this growing structure, and all the time the workers, like an army of ants, were swarming up and down ropes and all over the fragile structure itself. When the R100 was finished she had only two feet to spare at each end of her shed – quite a feat of engineering!

It took two years of design calculations followed by

two years of actual building before the two giants were ready to be tested. It also took 200 employees from Cardington, 150 airmen and 50 of the Bedford unemployed to walk the great R101 out of her shed for the very first time on October 12th, 1929. All the people who were in any way connected with her, together with friends and relations, turned up to give her a big cheer as she emerged from her chrysalis, glistening a pale, silvery pink in the early-morning sun.

Her first trial flight was over Bedford in good weather and lasted five hours, but as only two of the five engines worked properly it was not as impressive as had been hoped. On October 18th with all engines working properly she flew for nine and a half hours. Before she could improve on this a storm was forecast and she had to be hustled back into her shed and kept there until a sufficiently large number of people could be gathered to walk her out again, weather conditions permitting, of course. On November 17th she made a splendid flight of thirty hours soon after which she returned to her shed and stayed there until the following summer. Plans were now being set in motion for two long and impressive maiden flights for the finished airships. The R100 was to fly to Canada and the R101 to India. The Air Ministry announced that their own ship would be taking the new Indian Viceroy, Lord Thomson, to see his new empire in September 1930.

By this time however, quite unknown to the public, a large flaw had appeared in the R101. At the planning stage it had been decided that if the airships were to undertake transcontinental flights neither of them should weigh more than 90 tons without fuel. In this way the 'useful lift', ie, the total weight of passengers, fuel, luggage, stores, etc, would be 60 tons. (When you consider that fuel alone for a trip to India would weigh 25 tons, then 60 tons is really not very much.) After completion the R100 had a 'useful lift' of 54 tons but somehow the R101 could only manage 35 tons. Setting aside 25 tons of this for fuel, this left only 10 tons to fit in passengers, crew, provisions, etc. The R101's journey to India was quite impossible! This was a very embarrassing situation for the Air Ministry in view of the vast sums of money that had been poured into this much-advertised marvel. The beauty of her streamlined shape and all her clever, long-range devices would be useless if she could not carry passengers even as far as the coast. Obviously drastic steps had to be taken to improve her lift.

To begin with it was decided that if the wires supporting the gasbags were loosened, more hydrogen could be pumped into them, possibly giving as much as six tons of extra lift. This was done and showed a considerable saving, yet still not nearly enough. Finally it was decided that the R101 should be cut open and an extra section inserted in her belly large enough to carry more gasbags. This should give her a further nine-ton lift. However in order that the British public could see what they had paid for, she was to be shown off at the Royal Air Force Display in June, so her operation was postponed until after that. She was brought out of her shed a week before the display to be got ready.

It was a warm, fine morning and she was an imposing

figure, flying from the masthead. However, as people watched, a great tear, widening to 146 feet, appeared in her canvas cover. A team of men was sent up with needles and thread to mend it but the next day another large tear appeared in a different place. Clearly the whole fabric was completely rotten and would have to be replaced. This obviously could not be done before the display so the canvas was again sewn up and everyone crossed their fingers! On the morning of June 27th, covered in well-hidden patches, and with a crew praying for fine weather, the R101 flew off for a rehearsal at Hendon.

The display's high spot was to be when the R101 swooped gracefully down over the crowds and dipped in a royal salute. This had to be done at a rather low level, but certainly *not* as low as the 500 feet to which the airship suddenly dropped, almost scraping the top of the hangar! The commander immediately ordered her to gain height but as her nose came up there was a crack like a pistol shot – one of the main wire bridles had snapped. The journey back was difficult, even more so as the airship became heavier and heavier the longer she travelled and to get her home again 11 tons of ballast had to be thrown out – not a very happy omen for the people who had booked a passage to India!

The same sort of thing happened on the day of the actual display. The crew again found her extremely heavy and very difficult to handle but managed to put on a reasonable show for the quarter of a million people who came along to see her. No one suspected the truth. She was then returned to her shed to have extensive repairs to her leaky gasbags and to have her extra section fitted.

In the meantime her rival, the R100 flew off to Canada in triumph where she made a spectacular tour of the country and returned through very stormy weather without incident. She, at least, had proved her worth, yet when she locked back onto her mooring mast at Cardington there was hardly anyone there to welcome her. The ease with which the R100 had carried out this flight served to make the Air Ministry look even more ridiculous, especially as their own ship could not even fly to Hendon and back without getting into serious difficulties! They began to prod the overworked R101 team to get their ship off the ground as soon as possible. Lord Thomson had announced publicly that whatever happened the R101 must leave for India by October 4th and no more excuses would be tolerated. The airship people knew that if she was not ready to leave there would be no more money given for airship building and thousands of people would be out of work. The R101 just *had* to be ready.

The job of dividing the R101's hull had now begun. The operation was simpler and speedier than everyone had expected. The airship was simply sliced into two halves whilst hanging from the roof of her shed and the new section was quickly built and inserted in the gap. A whole new envelope was then fitted over the top. It was awkward to have to work in such a small space but as one team finished its stint of duty, another team took over and the airship was ready on September 25th for her long trip to India.

Yet it was still several days before the weather was considered calm enough for her to be brought out of her shed. By then it was October 1st and there was only time for one test flight for what was really a brand-new, untried airship.

On this test flight she was airborne for 16 hours but the failure of one engine made it impossible to try a full-speed test. So the R101 still had no Certificate of Airworthiness, without which she could not legally fly with a passenger load. Once more, in an attempt to save face, a 'temporary' Certificate of Airworthiness was handed over to the captain as the ship was being loaded. It had been decided, they said, that trials could be carried out with the passengers on board! An amazing gamble when people's lives were at stake.

At 6.30 PM on October 4th His Majesty's Airship R (for Rigid) 101, hung poised and ready for her maiden flight. The awesome sight of the world's largest airship, as long and as expensive as an ocean-going liner, bathed in the glare of a hundred searchlights, was worth the trip for the 3,000 people who had come to see her off. Excitement rippled through the crowd in waves, especially where there were throngs of Cardington people. After living with the R101 for six years and seeing her bring work and prosperity to the area, she had become a very important part of their lives and all willed her voyage to be a success.

There were to be fifty-four men carried on the R101 that fateful evening, six important passengers including the Secretary of State for Air (Lord Thomson) and the Director of Civil Aviation (Sir Sefton Brancker) and forty-eight crew led by the Captain (Flight-Lieutenant Irwin). Effort and expense had been poured into making the interior of the R101 as luxurious as possible. An historic dinner party for the Egyptian heads of state was to take place in Ismailia in the R101's dining-room and loads of specially-engraved silver and glassware had been taken on board for this occasion. There were many other 'vital' items like crates of champagne, barrels of beer and no less than twenty varieties of cheese for the enjoyment of the passengers. The item which caused the crew the most unease was the arrival of a pale-blue Axminster carpet which not only covered a lounge the size of a tennis court but also a 600-foot long corridor running almost the length of the airship. They knew that a layer of dust one-eighth of an inch thick covering the top of the airship would weigh one ton, so the weight of a carpet that size would make a big difference to their precious useful lift!

The beautiful white and gold lounge lined with potted palms was built to accommodate a hundred people and at the far end of it was a promenade deck where passengers could relax in deckchairs and admire the view. The R101 was also unique in having a smoking-room. Surrounded by five and a half million cubic feet of hydrogen, officers and passengers could sit in a small room of metal and asbestos and enjoy an after-dinner cigar. No one was actually allowed to take matches on board, however, and the lighters provided in the smoking-room were chained to the table so no one could take them away.

The great moment had arrived – the vast dirigible

pulled her nose away from the masthead cone and carefully backed away from it. Her bow, instead of soaring upwards as it should have done, dipped slightly and water ballast had to be dropped to prevent the great ship drifting even farther towards the ground – a gloomy portent! She began to increase the power in her engines. 'Look,' someone shouted, 'she's moving!' She was, but only just! By now heavy rain had begun to add its weight to the rest of her cargo and, as if this final load was almost more than she could bear, she lumbered slowly and sluggishly off into the night leaving behind 3,000 damp and dejected well-wishers.

It was a rough evening and the R101 only managed to travel at one-third of her expected speed, and at a rather low altitude, but the relief of actually getting off on time prevented the crew from feeling too concerned. After all, things were bound to improve when the rain stopped. Yet the rain seemed to have no intention of ever stopping, and it pounded unceasingly on the hull of the R101 as she reluctantly made her way towards the coast. In London the streets were thronged with people, even on such a dismal night, hoping to catch sight of the huge, new machine as she passed overhead. Some of them caught glimpses of her red and green navigation lights and the glow from her dining-room windows but most people heard only the noise of her engines churning miserably through the rain.

Inside the airship it was cheerful and warm and although the passengers barely knew each other at the outset, a few glasses of wine and an excellent dinner soon put them on friendly terms. As the great craft prepared to cross the sea at Hastings they retired to the smoking-room for an after-dinner cigar, feeling relaxed and contented. Yet there had been no improvement in behaviour of the R101 – quite the reverse in fact, since one of the after engines had now failed completely. Her height had dropped to a dangerous 700 feet and to the uneasy crew the white-capped waves below seemed uncomfortably close.

It took three hours to repair the faulty engine, a difficult task for Harry Leech, the engineer, whilst the airship was still in flight. It involved climbing up and down an outside ladder between the ship and the engine, something which most people would have found absolutely terrifying even in fine weather. Now, with the airship pitching and rolling in the grip of a storm, one slip on the wet rungs of the ladder and he would be in the sea in no time at all. He was delighted to get the faulty engine working again and be able to clamber back on board. The Channel crossing took two hours and as the airship crossed into Northern France the passengers retired to their cabins for the night, confident that all was well.

Unfortunately this air of confidence was not shared by the men in the control car. After leaving the sea, a fierce headwind met the R101 and slowed her already slow progress down to a crawl. They carried on in this unhappy manner until 2 AM when it was time for the crew to change-over watch duties. By then Harry Leech had just made a final tour of the engines, up and down the wet ladders in the pitch dark and had now made his way to the smoking-room to have a quiet cigarette before going to bed.

Leech relaxed into the comfort of a wicker easy-chair and put his feet up. Soon the combined warmth of the room and the comforting drone of the engines brought him to the edge of sleep. Suddenly, throwing him off balance, the airship dropped through the air like a lift with a snapped cable. It must have fallen at least a hundred feet and, before he had time to get up, the same thing happened again. This time the great ship was unable to right itself and continued along with its nose dipping at a steep angle, making it very difficult for Leech to get to his feet.

He knew, as well as anyone on board, that they were in great danger. Flying as near to the ground as they obviously were, it could only be a matter of minutes before they hit a church steeple or something worse! The airship dropped for a third and last time and, with a violent jerk, shuddered to a halt. As the lights went out Leech crawled uphill to the door and began to pull and twist at the handle. The crash must have buckled the metal because it was now impossible to open. Then came an ear-splitting roar that could only mean one thing. The five and a half million cubic feet of hydrogen had caught light. He was trapped in the middle of a raging inferno!

Seconds before this the R101 had passed over the French town of Beauvais and many of the sleeping residents had been awakened by the growl of the overworked engines struggling through the rain. Many of them rushed to their windows and saw the last of the R101 as she glided slowly and gracefully into a hill and exploded.

The wreck of the R101

Within seconds what had once been a proud airship was a mass of glowing girders and the bones of the R101 lay burnt and splintered under a foreign sky. In the seconds between the crash and the explosion five men had managed to leap clear, four engineers and a wireless operator. As they lay scorched in the heat of the burning wreck a sixth man, badly burned and bleeding, ran towards them. It was the foreman engineer Harry Leech who, driven almost mad by the heat in his asbestos prison, had managed to punch a hole through the wall, and without daring to hesitate for a second, had launched himself through it. He landed in a large, wet tree which obviously saved his life!

The six survivors were taken by French nuns to the local hospital where they were treated for burns and shock, and a message was sent at once to the Air Ministry in London. A party of experts flew at once to Beauvais. As they flew over her smoking skeleton the shocked observers could see nothing of the glorious creature that had flown off from Cardington only hours before. Her 777 feet of charred girders lay half in a field and half in a wood like a giant, torn cobweb and only the bizarre glimpse of a few golden pillars tossed on the ground reminded them of her past glory. Her dead lay inside the wood, covered by sheets brought by the villagers. Posies of flowers had been brought, and candles lit, by the sympathetic people who had done all they could to help.

No one really knows just why R101 met her death so quickly, but several reasons were put forward at the inquiry held afterwards. Some believed that part of her envelope had been ripped off and the gas cells punctured, others that the fitting of the extra bay had weakened her and that she had broken up before hitting the ground. Whatever the true reason for the disaster, the whole country was sickened by it and it sounded the death-knell for the British airship. The R100 was broken up and sold for scrap and the crew sent away to other duties. They were bitter men, grieved by the unnecessary death of their friends on the R101. The expensive mooring masts in Montreal, Ismailia and Karachi were dismantled and the metal sold for scrap. The giant mooring shed at Karachi, however, was more of a problem. It was so vast that for years a team of eighteen men had to be permanently employed in painting it because it was too near to other buildings to be blown up!

Carey Miller

Also written specially for Piccolo is the best-selling

## A Dictionary of Monsters and Mysterious Beasts 30p

Everyone is fascinated by monsters: mythical ones like the Minotaur or Werewolf, real ones like Tyrannosaurus Rex, just plain mysterious ones like the Abominable Snowman, or fictional ones like King Kong. An encyclopedic book of nearly 100 stories of strange beasts, with line drawings throughout.

## Submarines! 30p
Hazardous missions, daring exploits and rescue operations are retold in this book of true submarine stories.

## First Feats 30p
Peter Tunstall

Lindberg, Hillary, Magellan, Bell, Leonov – these men all achieved 'firsts' in their chosen field. 47 exciting stories are told in this lively anthology.

## Secrets of the Gypsies 25p
Kay Henwood

A lively account of the customs, rituals and magic of the Romany gypsy, that will fascinate children of eight and over.

## Bushrangers Bold! 20p
Frank Hatherley

True and thrilling stories of Australia's legendary outlaws.

## Piccolo All The Year Round Book 50p
Deborah Manley

For each month of the year, a wealth of ideas for things to make and do, facts about weather and history, famous birthdays, seasonal poems and much more. A superb 'year book' to dip into throughout the twelve months.

## Piccolo Book of Everyday Inventions 35p
Meredith Hooper

How were the cornflake, the typewriter, chewing gum or Coca-cola first invented? The stories behind these and many other indispensable items of our everyday life are vividly described in this excellently written and lively book.

You can buy these and other Piccolo books from booksellers and newsagents; or direct from the following address:

Pan Books, Cavaye Place, London SW10 9PG
Send purchase price plus 15p for the first book and 5p for each additional book, to allow for postage and packing

While every effort is made to keep prices low, it is sometimes necessary to increase prices at short notice. Pan Books reserve the right to show on covers new retail prices which may differ from those advertised in the text or elsewhere